Rethinking African Agriculture

Rethinking African Agriculture argues that rural communities in Africa are still shaped by non-agrarian factors both in livelihood strategy and social formation.

This volume renews and deepens research on the African peasantry by offering a fresh perspective drawn from the hitherto largely unknown Japanese research on the subject. The ethnographic fieldwork focuses not only on the microenvironment of the producers but also the broader historical context in which they live and work. The contributors argue that, in comparison with other regions of the world, Africa has never passed through an agrarian revolution that would effectively change the mode of production from within. Modernization efforts from outside have fallen far short of the ambition to transform agriculture in Africa. Rural Africa is still largely a natural society characterized by "non-agrarian" features, as evident in people's livelihood, social organization, and farming systems.

This book will be of interest to social scientists and anthropologists focusing on African development, agriculture, and agrarian societies.

Goran Hyden is a Distinguished Professor Emeritus at the Department of Political Science, University of Florida, USA.

Kazuhiko Sugimura is a Professor at the Center of Arts and Sciences at Fukui Prefectural University, Japan.

Tadasu Tsuruta is a Professor in the Faculty of Agriculture at Kindai University, Japan.

Routledge Contemporary Africa Series

Urban Planning in Rapidly Growing Cities
Developing Addis Ababa
Mintesnot G. Woldeamanuel

Regional Development Poles and the Transformation of African Economies
Benaiah Yongo-Bure

Nature, Environment and Activism in Nigerian Literature
Sule E. Egya

Corporate Social Responsibility and Law in Africa
Theories, Issues and Practices
Nojeem A. Amodu

Greening Industrialization in Sub-Saharan Africa
Ralph Luken and Edward Clarence-Smith

Health and Care in Old Age in Africa
Edited by Pranitha Maharaj

Rethinking African Agriculture
How Non-Agrarian Factors Shape Peasant Livelihoods
Edited by Goran Hyden, Kazuhiko Sugimura and Tadasu Tsuruta

Toward an Animist Reading of Postcolonial Trauma Literature
Reading Beyond the Single Subject
Jay Rajiva

For more information about this series, please visit: www.routledge.com/Routledge-Contemporary-Africa/book-series/RCAFR

Rethinking African Agriculture

How Non-Agrarian Factors Shape Peasant Livelihoods

Edited by Goran Hyden, Kazuhiko Sugimura and Tadasu Tsuruta

Routledge
Taylor & Francis Group
LONDON AND NEW YORK

First published 2020
by Routledge
2 Park Square, Milton Park, Abingdon, Oxon OX14 4RN

and by Routledge
52 Vanderbilt Avenue, New York, NY 10017

Routledge is an imprint of the Taylor & Francis Group, an informa business

British Library Cataloguing-in-Publication Data
A catalogue record for this book is available from the British Library

Library of Congress Cataloging-in-Publication Data
A catalog record has been requested for this book

ISBN: 978-1-138-61060-6 (hbk)
ISBN: 978-0-429-46568-0 (ebk)

Typeset in Baskerville
by Wearset Ltd, Boldon, Tyne and Wear

Contents

Acknowledgments

The present volume is the outcome of a series of conferences on "African moral economy" organized by the editors and local collaborators in Tanzania and Japan between 2004 and 2016. These meetings were concerned not only with the changing moral economy in contemporary village communities in Africa, but also with the historical uniqueness of sub-Saharan Africa as compared to other regions. We are grateful to the participants of the conferences, particularly Isaria N. Kimambo, Sam Maghimbi, Davis Mwamfupe, Deogratias F. Rutatora, Deborah Bryceson, Marc Humbert, Porphant Ouyyanont, Kumiko Sakamoto, Makiko Sakai, Yoshihito Shimada, Hiroshi Matsuda, Fumihiko Tsumura, Makoto Kuroda, Keiichiro Matsumura and Soichiro Shiraishi.

Our work has two intellectual origins in Kyoto. We would like to thank our precursors and colleagues from Kyoto University Graduate School of Agriculture, who provided a basic framework for the present research as part of their critique of modern agricultural science. We are particularly grateful to the late Keiichi Sakamoto, the late Toshinao Yoneyama, Osamu Soda, Tatsuro Suehara, Koichi Ikegami and Motoki Akitsu. This book is also based on the rich and unparalleled tradition of livelihood research among ecological anthropologists in Kyoto. Above all, the late Makoto Kakeya, as the leading figure in the research on the peasant cultivators in Africa, provided us with enduring intellectual and moral support.

We are also indebted to numerous individuals who have helped us in our field research. We particularly thank Nobuko Tsubaki and Casmir Nzullunge in Tanzania; Raphael Kinyanjui Wanjogu and Moses Kareithi Mwangi in Kenya; Godefroy Ngima and Jean Paul Kabeyene in Cameroon; Mohamed Yousouf ag Baye in Burkina Faso; Simon Awala and Ottilie Shivolo in Namibia; Maitseo Bolaane in Botswana; Chief Luchembe, Basheri, Male Mulenga and Margaret Bwalya in Zambia.

These fieldworks, along with the aforementioned international conferences, were supported by research grants from the Japan Society for the Promotion of Science (JSPS) known as KAKENHI in Japanese abbreviation, specifically JSPS KAKENHI Grant Numbers JP15255018, JP22255011, JP263 00014, JP17H04628, JP 08041059, JP11691186, JP20320131, JP25284171,

JP18H00776, JP17251002, JP15K16588, JP15H02598, JP16K21411, JP16 K16656, JP19K20537, JP18H02287, JP25871064, JP17K12970, JP18H03441, along with Grant-in-Aid for JSPS Fellows 24/8176. Our research is also based on the work conducted in the joint research project "Rethinking History of Agriculture and Rural Society in Africa [1] and [2]," sponsored by ILCAA (Research Institute for Languages and Cultures of Asia and Africa), Tokyo University of Foreign Studies. We are grateful to the members of this project, especially Naoki Ishikawa and Kaori Komatsu, who coordinated the project and gave us invaluable suggestions.

Finally, we would like to thank Larry Dodds, who checked the original manuscripts and our two editors at Routledge, Leanne Hinves and Henry Strang, for facilitating the production of our volume.

The Editors

1 Peasant agriculture in Africa

Science meets history

Goran Hyden

Introduction

Agriculture is at the center of academic and policy discourse on African development and has been ever since the colonial powers took over the levers of change on the continent. Regardless of approach, the African producers on the land have been overwhelmingly viewed as a problem of development. The main policy prescription has been that they must be dragged or lured out of their backwardness. They must become more scientific, specializing in agriculture instead of tilling the land without the explicit purpose of raising its productivity.

Despite its centrality and all the efforts to transform it, African agriculture has been in decline. From 1960 to 2005, the grain yield increased by only 60 percent. In comparison, the grain yield in East Asia increased by 220 percent in the same period. Africa's food production per capita declined by 15 percent from 1960 to 2005 and Africa's share in world agricultural exports also fell from 15 to 3 percent in this period. Africa went from being a net exporter of agricultural goods in the early 1960s to a net importer of such goods in the 2000s, spending, for example, US$33 billion on agricultural imports in 2006 (Government Spending Watch 2015). In addition to widespread degradation of the soil and natural vegetation, during this same period African governments and donors scaled back expenditures on agriculture. Despite a high-level decision by these governments at a Heads of State meeting in 2003 to reverse this trend and allocate 10 percent of the national budget to agriculture, only nine out of 47 countries had met this spending target in 2015 (ibid.). With such facts at hand, the verdict of the efforts over all these years to develop African agriculture into a scientific venture is that they rest not only on heavy-handed implementation but also – and above all – on an inadequate diagnosis. The search for answers, therefore, continues. One line of inquiry that has gained strength in recent years focuses on what one observer has labelled "working with the grain" (Levy 2013) which above all, implies understanding the context in which local and indigenous initiatives can flourish. In such an approach, history matters.

This chapter will examine how the scientific development lens has been applied to justify an onslaught on indigenous forms of production in the name of progress and productivity. It will show how approaches have evolved and varied over time but kept constant the notion that solutions must be found in outside interventions. This dominant view has been challenged by an alternative discourse that emphasizes the untapped and promising nature of indigenous forms of cultivation. This line of thinking, however, has prevailed in the margins of the official discourse and has only intermittently and in increments gained recognition in official policy circles. The chapter will discuss the evolution of these ideas and highlight why they may reach unprecedented influence in the emerging policy climate today and why attention to a hitherto little-known research tradition in Japan adds special value to this volume. Having set the stage for the rest of the book, it ends by offering an overview of the chapters to come.

The hegemonic development paradigm

Cognitive frames help explain how we view the world and the steps we take to make our livelihoods better and more sustainable. There are many such rivalling frames to make sense of the complex environment in which humans live and work. They vary in outlook for a variety of reasons. Some seek their origin in nature, others in heaven. None, however, has been more influential in recent generations than the paradigm based on science. Although the foundation of science has been laid by many civilizations stretching over thousands of years it has become the prerogative of the western mindset in the last few centuries. Its core notions have been to control and conquer, initially the physical environment, but eventually also people. The modernization theory which has underpinned international development practice since the mid-twentieth century is a benign version of this mindset. It replaced the cruder – and more brutal – version that led to the colonization of Africa a century earlier. In fact, today science is with few exceptions portrayed as positive, progressive and beneficial because it offers a view of the world as lawful and because the same laws are assumed to hold everywhere it is viewed as providing the key to unlocking human misery. The simple and neat interpretation expressed in terms of everyday mathematics has gained prominence not only in the natural and physical sciences but equally so in the medical and social sciences. This reductionist form of science has permeated even development practice and thus helped set a universal agenda for approaching problem-solving, African agriculture being no exception.

The paradigm that has so extensively permeated thinking about development in the last 70 years is not a monolithic creation but reflects the evolution of ideas over time. Lessons have been learned and policies have been changed but one variable has remained constant: modernity! This western notion has provided universal justification for replacing forms – whether material or mental – that fail to generate the most profitable or productive results. This constancy notwithstanding, the path has not been linear. A look back to the years immediately after the end of the Second World War reveals the twists and turns that have shaped the parameters of the dominant development paradigm.

It was invented in response to the challenges posed by the decision during the Truman Administration in the United States and eventually by other nations to launch large-scale foreign aid not only for reconstruction in postwar Europe but also for overcoming poverty in other regions of the world. There was no real practical experience to work from and finding an overarching theory to provide justification for this type of engagement became an immediate priority. Inspiration was initially sought in the western intellectual tradition of thinking about the nature of social change (Shils 1963: 11–12). Two academic publications became especially influential: (1) Almond and Coleman (1960) who developed a structural-functionalist approach which argued that functions are the same in any political system and differences can be explained by their degree of structural differentiation; and (2) Easton (1965) who treated the political system as a subunit of a wider societal system with its own boundaries. These authors not only started a revolution within the discipline of political science, but they also paved the way for a view of development as a process whereby the western experience becomes the model for the rest of the world and science its prime vehicle.

These theorists viewed the political system as an organism in which all parts are connected through their own feedback loops. This was a totally new way of interpreting development; but because it did not easily lend itself to operational measurements, it soon lost out to a more mechanistic view of politics – one that presupposed each part of the political system could be analyzed – and changed – independently. Apart from being called into question because of their high level of abstraction and difficulty to operationalize, these early attempts to lay the foundation for comparative development analysis also generated other forms of critique. One line questioned the tendency to treat social change as a cluster of internally compatible variables that "keep the system going" regardless of challenges to its legitimacy. As Huntington (1965) noted, if one unpacks the concept and treats selective aspects of it, e.g., institutions, one finds that modernization may strengthen instead of weakening traditional institutions and values, and rapid social change in one sphere may serve only to inhibit change in another. A second line of criticism directed attention to the notion that tradition and modernity represent two mutually exclusive, functionally interdependent clusters of attributes. Rudolph and Rudolph (1967) showed in their research in India that in many instances "traditional" institutions and values may facilitate rather than impede social change. Modernization, therefore, cannot be equated simply with the destruction of tradition because the latter is not a prerequisite of modernization (Tipps 1973).

Nisbet (1969: 190–1) summarized the problems with this first attempt at comparative analysis by writing that it

> is hardly more than a shoring-up of the idea of progressive development generally and, more particularly, of the belief that the recent history of the West could be taken as evidence of the direction in which all mankind *would* move, and flowing from this, *should* move.

A prominent African scholar, Ali Mazrui (1968: 82), referred to the reassertion of these ideas of social evolution as "the self-confidence of ethnocentric achievement."

The initial modernization paradigm lasted through the first decade of independence but by the mid-1970s, the battleground had shifted to economics. The belief in modernity remained strong but the choice now stood between two rivalling economic models – one based on Marxian or socialist premises, the other on the long intellectual tradition in western thinking dating back to Adam Smith. In his liberal market theory, individuals are perceived as always ready to maximize their own utility. This simple formula gradually emerged as dominant and by 1980 had become policy practice throughout Africa. Treated as the core of a neoliberal economic ideology, it is in contrast to the initial paradigm by stressing the competitive rather than the cooperative nature of human behavior. The injection of rational choice spurred the growth of game theory and other similar models to analyze development policy. The landmark book by Elinor Ostrom, *Governing the Commons* (1990), not only earned her a Nobel Prize but also became dominant as a way of demonstrating the rationality of all actors and their ability to design and reform institutions that set the parameters for their actions. The neoliberal model of development met with its own critique along two lines. The first maintained that parsimony tended to become an end in and of itself overlooking how structures and institutions confine agency (e.g., Young 1994 and Herbst 2000). The legacy of the colonial state could not be overlooked which meant that bureaucratic influences would extensively modify the basic market principles. The other stemmed from an acknowledgment of the rich literature on the communal feature of African society and its implications for human behavior (e.g., Ekeh 1975; Wai 1978; Hyden 1980).

The third incarnation of thought about the inevitability of a western route to progress elevates democratic political institutions to the role of key drivers of change. Pursued under the mantra of "good governance," western donors insisted that their foreign aid be tied to acceptance of western democratic values. Modernity here builds on yet another intellectual tradition in western thinking – this time grounded in John Locke's notion of how people can govern themselves in a free and fair manner. Like the Structural Adjustment Programs of the 1980s, the Good Governance agenda of the 1990s reinforced a policy practice based on externally driven reform packages. The emphasis was on funding public sector reforms in African countries. One of the first offshoots from this line of thinking was the New Public Management approach, which argued for a public service driven by criteria previously associated primarily with private sector operations, such as efficiency, results-orientation and a delegation of operational authority to non-state actors. In the early 2000s, studies showed that these efforts easily fall short of their objectives largely because they fail to pay attention to underlying cultural values. As for example, Andrews (2013) has demonstrated: it is easy to reform organizational structures but much more difficult to change the institutional dimensions that drive human behavior inside these structures.

Moments of revisionist influence

Such critique is part of the revisionist thinking that has evolved in parallel with the dominant paradigm and been perpetuated in academic circles and in the policy practice of non-governmental organizations (NGOs). Its influence on official policy has been intermittent, but not inconsequential. Three such moments deserve mentioning here.

The first occurred in the early years of independence before the donors had found a common platform for providing aid. During the 1960s and into the following decade modernization was not so much a set agenda as a bundle of interventions carried out in response to government preferences in Africa's newly independent countries. These governments generally shared the same development objectives as the donors and by that time there was little or no critical awareness of how these objectives could be most sustainably put into practice. There was political space for bold initiatives such as *ujamaa* – a form of communal village production on land – in Tanzania and in the absence of a common agenda, donors found themselves supporting such ventures regardless of whether they adhered to their normative preferences.

The example of *ujamaa* is instructive because its socialist content was out of line with what donors would have preferred but it was generally assumed that rural development in Africa must entail creating a viable institutional infrastructure for the provision of social services. This was viewed as an especially important intervention in a country like Tanzania where most people lived in scattered homesteads. The Government of Tanzania, like the principal donors, notably the World Bank, shared the idea that placing people in villages is a prerequisite of development. During colonial rule the British had used settlement schemes as a way of modernizing rural life albeit with moderate success. The World Bank on the eve of independence had tested a similar approach. It turned out that these Bank-funded schemes attracted settlers from already densely populated areas rather than from regions dominated by scattered households. In addition to missing the most critical category of people, the projects proved quite expensive. In the immediate post-independence period, therefore, a consensus about the need for villagization persisted but there was still no agreement about how to achieve it in a sustainable manner. President Nyerere's declaration in 1967 that the country would adopt socialism, and as part of it create villages in which households can cultivate land communally while still retaining a private lot for their own subsistence needs, seemed to provide the long-sought answer. It certainly became a catalyst for extensive outside funding of rural development in the country throughout the 1970s. At a time when donor governments trusted Africa's new leaders to have the commitment, if not all the capacity, to achieve positive development outcomes, they lined up behind Nyerere's policies without first questioning their institutional content. Doubts began to arise only after it became obvious that Tanzanian peasants, whether in villages or not, found ways of circumventing the bureaucratic implementation machinery.

The emerging critique tended to blame party and government bureaucrats for screwing up the implementation of the policy (von Freyhold 1979 and Coulson 1982). A parallel critique queried the extent to which Nyerere's *ujamaa* was truly socialist (Cliffe and Saul 1973). This mainstream analysis ignored the possibility of other interpretations. The question was not really whether Nyerere's ideas were Marxist enough. It was rather that it was propelled by a desire other than the wish to control nature. His philosophical foundation was much more in tune with Oriental, especially neo-Confucian thinking which tries to discover the ethical principles that stem from humanity's intrinsic connection to nature. It was not so much about increasing productivity on the land as previous settlement schemes had attempted but rather about discovering and realizing the moral principles that Nyerere perceived as inherent in African societies. In this respect *ujamaa* was not an effort to adopt but to reverse the western rationalist thinking that had dominated during the colonial period and was being reasserted by both western liberal and neo-Marxist thinking after independence. Other African leaders tried the same reversal of values – President Kaunda's Humanism in Zambia being another example. Yet others did so without being as articulate in their ambition as Nyerere or Kaunda. The revisionist moment after independence, however, did not last long. Because African governments lacked enough capacity of their own, donors could close the window on alternative forms of development and did so successfully in the late 1970s by becoming more insistent that Africa's new states adopt economic policies that reflect western forms of rationalist thinking. The inevitable results were neoliberal economic reforms.

The second moment of revisionist influence occurred not long after the first had come to an end. Unlike the first which entailed a broad-based social transformation propelled by the wish to recreate values that were perceived as lost in the colonial period, the second moment was more directly focused on changing agricultural cultivation practices. Beginning in the early 1980s, it centered on the introduction of a farming systems research (FSR) approach aimed at tailoring farm innovation to the household realities in which peasants produce and live. An influential source at the time was the book by Paul Richards on indigenous agricultural revolutions in Africa (Richards 1985). FSR was advanced as a way of increasing the content of indigenous knowledge in farming innovation but it proved difficult for researchers and extension services to move outside the western knowledge system (den Biggelaar 1991). Advocates of indigenous knowledge and practices became captives of a discourse that focused more on environmental protection than on raising agricultural productivity. As international donors bought into this narrative and abandoned support of agriculture in favor of environmental issues, whatever remained of the call for indigeneity – further developed for practitioners by Robert Chambers (1983) and his colleague, Gordon Conway (Chambers and Conway 1992) – was adopted by NGOs working with poor farming communities around the world. Support for agriculture in Africa has never returned to the top of the international donor agenda. Interest has shifted to other concerns, notably social development sectors like education and health. Investments in agriculture have been left to the private sector as donor governments have adopted the view that peasant agriculture is not going to solve Africa's food deficits.

The third moment of challenges to the dominant paradigm has gradually emerged from the rethinking of what drives change and development. Throughout the years of Structural Adjustment and Good Governance, donors took for granted that the "right" policies and "right institutions" drive development. Revisionist theories may have agreed with the diagnosis that the problem is political but offered very different aid policy prescriptions. Two of the most influential books contributing to an alternative interpretation of how to build institutional capacity in developing countries were written by Acemoglu and Robinson (2006) and North, Wallis and Weingast (2009).

These authors have developed a framework which demonstrates under what conditions elites would be ready to share power with non-elites. Arguing that elites are not a unified group but engage in competition with each other – and even go to war with each other – they are primarily engaged in bargaining among themselves rather than sharing power with others. To highlight the challenges associated with the historical transition to democratic forms of governance, North and his colleagues differentiate between social orders with open access to both political and economic activities, on the one hand, and those, on the other, that only provide limited access. The latter type also labelled 'natural states' exists on a continuum ranging from fragile states, characterized by political instability and violence, at one end, to mature natural states such as emerging markets, on the other. Few countries outside Western Europe,

North America and some former British colonies have evolved beyond this stage, and natural states comprise approximately 95 percent of countries.

Based on extensive historical analysis, they explain that the transition from authoritarian to democratic structures must be consistent with the logic of the natural state. To trigger a transition, it is important to create preconditions that put the elites in a position where it is in their own interest to transform personal and privileged deals among themselves into impersonal laws, so that all members of the elite groups are treated the same way. Change does not come about merely through the supply of aid money but stems from incentives that lead to a demand for it. Political will is a precondition for institutional reform.

An important contribution to this framework is the theory of how societies reduce or control the problem of violence. In natural states the control over violence is disbursed among many groups, which therefore tends to make them fragile and prone to civil wars. Natural states, however, have their own way of solving the problem of violence through rent creation, granting powerful individuals and groups privileges so that they have incentives to cooperate rather than fight. Natural states cannot be viewed as merely dysfunctional. They have their own political logic. Although they are less robust to shocks than open access orders, they generate internal forces that provide for two of the basic tasks of all societies: stability and order. Natural states may appear to be corrupt according to the norms and values of open access orders, but corruption and rent seeking by elites is an inherent part of the operation of this social order. Combating it therefore entails changing the social order altogether. As Khan (2017) also argues, failure to understand how the much more visible and direct connections among political, economic, religious and military privileges are integral to the social order is a major impediment to a better development policy.

Following a growing frustration with their ambition to change Africa by insisting on their own normative models, donors began to engage in some measure of self-reflection. This new thinking has emerged in the wake of the 2005 Paris Declaration on Aid Effectiveness and was confirmed in the 2011 Busan Partnership Statement signed by Heads of State from both developed and developing countries. The question is still how seriously this initiative will be taken and how far it might be implemented. It has been boosted by influential academic writers who argue for the need to "work with the grain" (Levy 2013) and acknowledging "political settlements" as key drivers of policy change (Khan 2017; Booth and Gooloba-Mutebi 2012).

The relevance of peasant studies

This most recent twist in thinking about development occurs at a time when there is also a renewed interest in peasant studies. As a focus of study, peasants in the past hundred years have featured on and off in the literature on social change and development. It began with studies of the conditions in Europe and was later extended to other regions. The definition of "peasant" has varied but the underlying notion has typically been a reference to groups of people at the margins of society. They were portrayed as producers on the land subject to legal and political restrictions, notably the right to mobility (Dalton 1972; Shanin 1973). Similar ties of bondage could be found in China and other Asian societies (Moore 1966) as well as in Latin America (Huizer 1973). Thus, definitions tended to include references to their subordinate role in serving other social classes, whether described as landlords or townsfolk. Shanin provided a general definition that was commonly used by other social scientists at the time. He defined peasantry as having four essential and interlinked facets:

> ... [t]he family farm as the basic multi-functional unit of social organization, land husbandry and usually animal rearing as the main means of livelihood, a specific traditional culture closely linked with the way of life of small rural communities and multi-directional subjection to powerful outsiders.
>
> (Shanin 1973: 64)

Yet, questions remained about whom to include in the peasantry. The tendency was to use the concept not only with reference to cultivators of land but also to fisherfolk, artisans, pastoralists, even small-scale miners. This broad categorization also led scholars to argue that the peasantry is not homogenous. Eric Wolf (1966), for example, had sought to delineate social structural types based on whether they had secure land rights or, alternatively, were tenants, sharecroppers or resident laborers on large properties. Furthermore, 'peasants' tended to be distinguished from 'farmers', since the former were viewed as aiming at subsistence and producing cash crops primarily for survival and thus being more interested in maintaining their social status rather than investing and expanding the scale of their operations, as was allegedly the case with the latter.

The debates of the 1970s were heavily influenced by neo-Marxist theories about the proletarianization of the peasantry. The regional focus was on countries in Latin America and Asia as confirmed in the useful review of the content of the once defunct *Journal of Peasant Studies*[1] published in the initial issue of its successor – the *Journal of Agrarian Change* (Bernstein and Byres 2001). Apart from studies of the effects on the peasants of settler and plantation agriculture in southern Africa, the continent was much less present in these debates, the main exception being references to Marxist anthropologists in France like Claude Meillassoux (1975). Although he and his Francophone colleagues tended to acknowledge the significance of Africa's precolonial modes of production for the analysis of contemporary society, the debates in Anglo-Saxon publications continued to be based on the premise that the African peasantry was undergoing the same process of exploitation as their counterparts in other regions of the world. The African peasantry had no choice but to give in to the historical forces that confine it to the margins of society and leave it as a vanishing resource.

This rather somber portrait stands in contrast with the conceptualization that has emerged in recent years according to which the world is witnessing a restoration of many of the core principles of peasant production, notably: (1) diversification of income to enhance security; (2) alignment of production with environmental conservation; and (3) reduction of dependence on volatile market exchanges. This is the main argument in van der Ploeg's book on the 'new peasantries' (van der Ploeg 2008, 2018). Based on research in Italy, the Netherlands and Peru, he describes a process that occurs in countries where industrialization and commercialization have left consumers dependent on corporations for their food security. There is a moral dimension to van der Ploeg's analysis somewhat resembling James Scott's portrait of farmers in Southeast Asia as capable of undermining or withstanding the forces of commercialization (Scott 1976).

This "new peasantry" argument rings a familiar bell in places where family farming has been marginalized by land consolidation and industrial forms of production. It captures the reaction in recent decades to this process and points to a growth in measures such as direct marketing, landscape management and on-farm processing as evidence. It is less immediately applicable, however, to the African conditions where nuclear households are still the exception and consumption rather than production tends to be the foremost livelihood determinant. This observation does not imply a rejection of the general thesis that peasants are a resource and not a problem but it considers the point made by Alexander Chayanov a century ago (albeit reproduced in English only much later, e.g., Chayanov 1966) that it is not rational for peasant producers to work harder and longer than necessary to satisfy a locally acceptable standard of consumption. The Russian peasantry – the principal focus of his analysis – like African peasants today lived a life which was only marginally dependent on what was happening in the macroeconomic environment. Chayanov, unlike his Leninist contemporaries, doubted their willingness to voluntarily adjust to a socialist system demanding an increased surplus production for collective purposes.

Chayanov becomes relevant to development policy thinking today because of the call for greater sensitivity to context. Blueprints are increasingly being questioned in favor of analyses that consider what might work from the ground up. This includes consideration of the fact that in a historical perspective, conditions across regions and countries may vary. For example, peasants in Africa are not faced with the same type of social transformation as their counterparts in Asia and Latin America (leave alone Europe). To be helpful, the historical perspective must go beyond the conventional time frames applied when analyzing Africa, e.g., colonial and postcolonial. It needs to incorporate a Big History view which recognizes that unlike other regions of the world, Africa has not yet gone through an Agrarian, leave alone Industrial Revolution. Peasant agriculture in Africa is not a specialized occupation. It is a livelihood that relies on diversification and follows a sharing rather than production logic associated with a Natural Society model that historically precedes the transition to Agrarian Society. In the wider conceptualization of global development, the rather exceptional features of rural life make Africa its own region – the South of the Global South. Modern science does not make sense there without first being placed in a Big History context.

The Japanese research tradition

This volume is a contribution to a better understanding of Africa's special development challenges and opportunities as seen from the bottom up. It acknowledges that peasants in Africa are autonomous enough to shape their own livelihoods while interacting with nature and other social groups. Although there are peasant producers whose livelihoods are extensively integrated into the modern market and thus view investments in agricultural productivity as fully rational, they are not typical. They can be found on fertile soils notably in the highland areas of eastern and southern Africa. The prosperous coffee farmers on the slopes of Mt Kenya and Mt Kilimanjaro, however, are the exception, not the rule. Most peasant producers in Africa pursue their livelihoods in less inviting circumstances that call for a more diverse and flexible approach. Their dependence on state and market is more marginal. As a result, microeconomic considerations are as relevant for development policy analysis as macroeconomic factors. This volume is an attempt to fill out the contours of this complex political economy that involves the coexistence of precolonial, colonial and postcolonial features. It offers a perspective that is fresh and intellectually challenging for academics and policy practitioners alike. It is also special because it draws on a long research tradition in Japan that speaks to understanding context and local ownership, but which has been largely unknown in western – and African – literature on the continent's agriculture.

African studies in Japan which originated in the 1950s, more specifically at Kyoto University, grew out of a broader interest in the evolution of humankind. It is more fully introduced in Chapter 2. Suffice it to mention here that the core argument of this research tradition is that formulas for agricultural development based on the experiences of Agrarian Societies with their agricultural specialization are misleading and produce questionable and unsustainable policy prescriptions. Rural communities in Africa are not specialized but encompass many non-agrarian elements, for instance, pastoralism, hunting and gathering, and other diversified subsistence as well as income-generating activities. This is in contrast with rural communities in the Old World which have experienced an Agrarian Revolution.

The main theme of this book is to explore the non-agrarian features of the African peasantry, which stems from our assumption that the major parts of sub-Saharan Africa have never passed through an "agrarian" stage of historical development. In line with the suggestion by Shumpei Ueyama (1966), we call this pre-agrarian existence "the Natural Society stage", its main features being developed and discussed in Chapter 2. We view the omission of these features as a major reason why modernization of African agriculture has been unsuccessful. As the various contributions to this volume illustrate, Natural Society legacies are still very much part of rural life and have a direct influence on how people approach agriculture. With a growing migration of rural residents to urban locations, these legacies are also implanted into life in the towns.

The concept of Natural Society entails, in the first place, that human production systems are embedded in nature, as can be seen most directly in traditional hunter-gatherer societies. This embeddedness in nature, however, is present much more widely. Most notably, smallholder peasant producers have also attuned their livelihood to local ecology, constantly adapting themselves to the shifting environment through migration, through changes in farming systems, and/or through the incorporation of other subsistence strategies such as hunting, gathering and livestock herding. Their livelihood pattern, therefore, lacks uniformity and adopts different crop regimes or subsistence strategies from one place to another, even within the same locality. It is this diversity, multiplicity and mobility which have characterized agriculture and subsistence in rural Africa, in stark contrast to those in Agrarian Societies which are marked by uniformity and sedentariness.

Because most African societies never developed their own "agrarian states," which are based on effective exploitation of peasant surplus products, people are not caught in permanent structures reflecting the presence of a singularly powerful state or market. In Agrarian Societies, the state controls the production systems of the peasantry as a subordinate class, hence the uniform farming systems and livelihood patterns. In contrast, in the greater part of precolonial Africa, peasant cultivator communities have remained largely autonomous, dispersed, and without a social stratification based on individual land ownership. Their production patterns have been dictated primarily by the given natural environment and their own consumption patterns rather than state institutions. The assumption that consumption drives production is similar to Chayanov's argument but there is a significant difference: his theory was based on consumption within a single household while in rural Africa it is shared consumption among several interlinked households that affects production patterns. In Africa, a unit of production is not necessarily the same as the unit of consumption. Foods and other goods are shared and redistributed among different households, through which social relations are reproduced and their production level is determined. In this sense, the social relations of peasant cultivators in Africa are more strongly ruled by the logic of reproduction and consumption than by that of production.

The above claim is reminiscent of the argument about the "uncaptured peasantry" and the "economy of affection" that Hyden (1980) made in his original analysis of rural Africa. This book is an attempt to flesh out the thesis of an uncaptured peasantry using the ecological anthropology perspective of the Kyoto School. Although 40 years have passed since Hyden's original proposal, the legacies of a Natural Society are still very much present not only in rural Africa but increasingly also in its urban areas, the result of rural migrants moving to the cities carrying with them such key values as a diversified livelihood pattern and a readiness to share their wealth with relatives and friends. The informal economy of affection, as one example of a wider range of economies of sharing, which is discussed further in Chapter 3, is not only a significant factor in shaping social formations but also a hindrance to a form of capital accumulation for use in productive pursuits.

What makes the contribution by the Japanese researchers featured in this volume so interesting and relevant is their insistence on a full understanding of the complex historical conditions in which Africans pursue their livelihoods: why they remain relatively autonomous, why they rely on informal institutional mechanisms and why raising productivity on the land is not a priority. In this perspective, the peasantry in Africa is not "new." The point is rather that attention to its true historical conditions is long overdue. The observations and theoretical insights conveyed by the Japanese researchers make this volume an original contribution to African studies in general and peasant and agrarian studies in particular while at the same time raising critical issues with policy implications.

Organization of the book

The next two chapters help to further explain the main thesis of the volume and clarify its key concepts. In Chapter 2, Tadasu Tsuruta explains the distinction between Natural and Agrarian types of societies which is central to the Japanese approach of understanding rural Africa. Drawing on the works of philosophers and anthropologists based at Kyoto University and making references to cases discussed in subsequent chapters, he shows that although rural Africa has been increasingly touched by global forces, it remains extensively influenced and shaped by non-agrarian features of life. This may be most evident in the livelihoods of hunters and gatherers but can be found also among other groups, whether they are cultivators, pastoralists or engage in a combination of the two. Non-agrarian features include the ways peasants in Africa organize their livelihoods, their farming systems and their social exchanges.

In Chapter 3 Goran Hyden introduces the concept of "economy of affection" and discusses how it is a structural manifestation of Natural Society and its logic of sharing. He sees this logic permeating not only rural but also urban Africa as more and more people, especially the younger generation, migrate to the cities. They do not abandon the values and norms from home which include the obligation to share with family and kin as well as in certain circumstances with even larger communities of consumption. He shows how the economy of affection gives rise to a host of informal institutions that play an important role in development and governance of African countries. The chapter ends with a comparison with other forms of moral economy that have features prominently in the literature.

The next six chapters constitute the empirical core of the book and feature the research of a cross-section of Japanese researchers devoted to studying rural Africa. These empirical studies cover diverse livelihood patterns and ethnic groups in rural Africa (Figures 1.1 and 1.2).

In Chapter 4, Haruna Yatsuka and Kazunobu Ikeya discuss the lasting non-agrarian influences of hunters and gatherers in many rural parts of Africa. Not only do hunters-gatherers still exist in small pockets, living side by side with peasants and pastoralists, but there is also a sizable population

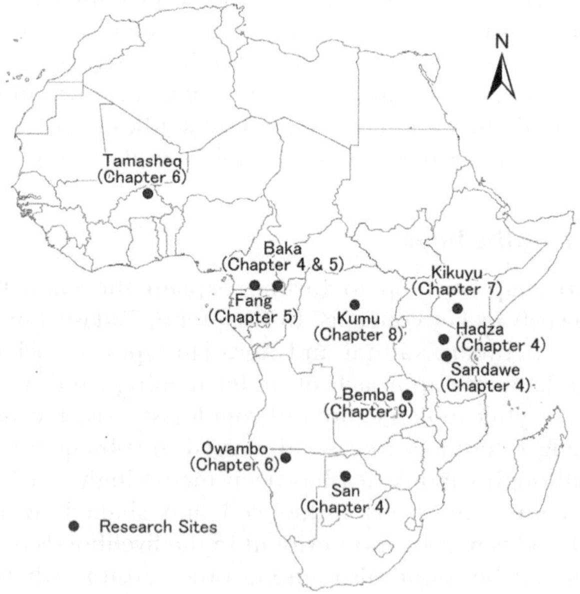

Figure 1.1 Study areas and ethnic groups.

Source: Author.

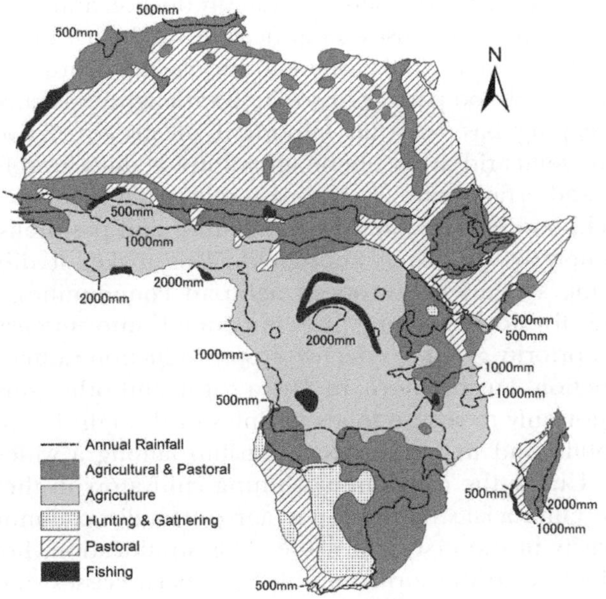

Figure 1.2 Types of subsistence economy in Africa.
Source: Adopted from Murdock (1959: 18).

of cultivators influenced by the hunting-and-gathering tradition both in their food culture and social institutions. These livelihood patterns are grounded in a philosophy that stresses diversification over specialization. Drawing mainly on local cases from Tanzania, Cameroon and Botswana, the authors explain the persistence of these non-agrarian features in ex-hunter-gatherers' practices of farming and sharing, and how Natural Society shapes their livelihood patterns and thus sets the parameters of social change.

Chapter 5 is a case study of the livelihood patterns among peasants engaged in shifting cultivation, more specifically a form of swidden agriculture, in the Cameroonian rain forest. Shifting cultivation has long been the major farming system in sub-Saharan Africa, especially in the rain forest and its periphery, the savanna woodlands. Using Kakeya's classic argument about the livelihood pattern and value system of shifting cultivators, Kenta Sakanashi, Kazuhiko Sugimura and Tadasu Tsuruta describe *first* how villagers utilize forest resources for their farming practices and other subsistence activities acting as 'generalists' of ecological management and not as 'specialized' farmers; and *second* how they have built a symbiotic relationship with neighboring hunters and gatherers to enrich their livelihoods. The chapter also explores the implications of these non-agrarian features for the introduction of new farming methods into the African rain forest with special reference to commercial cacao production and agroforestry.

Drawing mainly on intensive empirical research in Burkina Faso and Namibia, Yuichiro Fujioka, Yudai Ishimoto and Tadasu Tsuruta use Chapter 6 to discuss the livelihood patterns associated with agropastoralism in Africa. Livestock keeping has been an important means to survive the harsh climate of its semi-arid areas where agriculture is quite unreliable due to insufficient and erratic rainfall. Agropastoralism is a uniquely African way of rural livelihood in that extensive agriculture and pastoralism are practiced separately by the same household. This makes it different from mixed-farming systems in European agrarian communities, where agriculture and livestock keeping are integrated into intensive farming systems with priority given to effective crop production rather than to livestock production. Furthermore, in Africa cattle and other small stock are important not only to secure foods in times of drought but also as social wealth to build and maintain ties of kinship among a widely dispersed population. Unlike the case among shifting cultivators in the rain forest, there is a certain social stratification in agropastoralist communities owing to the disparity in ownership of cattle. This stratification, however, does not normally lead to the formation of rigid social classes since assets (in the form of animals) are typically dissipated or lost in one single generation either through social transaction or unpredictable change in environmental conditions. This chapter also pays attention to how wealth is redistributed among villagers in agropastoral communities through the economy of sharing built into their culture.

In Chapter 7, Noriko Ito and Tadasu Tsuruta discuss how moral economy considerations permeate even among commercialized rural producers, more specifically in Kenya's Mwea Irrigation Scheme. In the country's largest irrigation scheme, highly commercialized rice production has been promoted since the 1950s. This chapter shows how the original settlers engaged in market behavior to improve their livelihoods but how they were increasingly faced with how to accommodate the needs of their families, including the second-generation villagers who are generally short of grain. Members of the first generation, who were generally richer, tend to share rice with the second generation in need of help. These actions, that combine market and moral economy considerations, make the livelihood patterns in the scheme quite different from rice-growing communities in other parts of the world, where the Green Revolution had eliminated the former economy of sharing. This chapter then ends with the comparison of rice-growing communities in Southeast Asia with the case in Kenya, highlighting a crucial difference in how the villagers share the harvested crop among different households.

Kazuhiko Sugimura summarizes the arguments of preceding chapters in Chapter 8, drawing also on his own study conducted in the Congolese rain forest. It examines the lasting influences of Natural Society in all spheres of peasant life, but particularly in the sphere of consumption: how they organize their livelihoods in tune with nature rather than with market incentives or state policy; and how they engage in social exchanges where the value is peculiar to each case rather than set in conventional standard monetary terms. These examples are in contrast with what typically happens in Agrarian Societies where specialization leads to greater exploitation of the soil. Finally, this chapter argues for the necessity to establish a new agricultural science suited for Natural Societies, in place of modern agricultural sciences developed in Agrarian Societies, which have been characterized by mono-cropping, mixed farming, higher use of input, and individual ownership of farmland. Modern agricultural sciences have always resulted in a top-down approach to technological change, which has never worked well in rural Africa, in a stark contrast to grassroots innovation.

In Chapter 9 Yuko Sugiyama provides a rich historical account of grassroots innovation among the Bemba in northern Zambia. It highlights the social dynamics of the diffusion process of newly introduced techniques throughout the twentieth century. It begins with the description of the livelihood pattern of the Bemba people, who had long been dependent on shifting cultivation known as the *chitemene* system. By examining the process of introduction and diffusion of innovation in both on-farm and off-farm activities, this chapter shows how these new technologies were "socialized" and "localized" through the process of redistribution of foods and income-generating resources, leaving the door open for anyone who needs to gain access to new techniques and resources. It also argues that the innovation process in Bemba society is closely associated with its levelling mechanism and multifaceted nature of communication network and livelihood.

Chapter 10 contains the overall conclusions of the study. Written by the Editors it addresses such questions as (1) what is the contribution of Japanese scholarship on rural Africa? (2) What are its implications for comparative research on Africa? (3) How does this knowledge help inform current international development policy thinking? (4) Where may Africa be heading given current global structures? The discussion of these questions will be placed in the context of the emerging international development agenda to show how this volume addresses both challenges and opportunities related to the growing call for attention to context and local ownership of development initiatives.

Note

1 JPS was later revived under the editorship of Jun Borras, a Philippine activist-scholar, with a strong anti-capitalist overtone.

2 Missing out on the Agrarian Revolution

African peasantry in historical perspective

Tadasu Tsuruta

Introduction

This volume is based on a very simple premise: sub-Saharan Africa has never developed an "agrarian society," unlike countries in other continents where the peasant class was subjugated by despotic or feudal states to produce surpluses for the rulers. Most pre-modern village communities in Asia and Europe were based on a continuation of settled agricultural communities resulting in a high degree of incorporation into the higher administrative authorities of the state. In contrast, precolonial African rural societies were made up of relatively dispersed and self-sufficient communities which were not always specialized in agriculture. Landlordism was virtually non-existent; so was the presence of a landed class. Thus, Africa appears to be an exception to the general pattern present in other agrarian societies.

Almost 40 years ago, Goran Hyden called our attention to Africa's "uncaptured peasantry," a term which explains the situation where both state and market fail to involve peasants in modernization projects (Hyden 1980). Since then, the disparity in rural economic development between Africa and the rest of the world has grown even greater, emphasizing the anomaly of development in rural Africa. The "uncaptured" nature of the peasants and the resulting poverty and underdevelopment is related to the non-agrarian features of rural societies in Africa.

Some may frown upon our argument as "African exceptionalism" or a sweeping generalization. We believe, however, that we will never get a meaningful perspective on the future development of sub-Saharan Africa without first exploring the historical difference between Africa and other societies from a macroscopic view of the evolutional history of civilizations. Neo-Marxist scholars like Rodney (1972) and Amin (1976) have made attempts to unravel the root causes of African underdevelopment by blaming colonialism – and by extension capitalism – for the continent's underdevelopment. Their perspective, however, has been focused on demonstrating the similarities rather than differences between regions of the world. Others, like Henley (2012), have tried to compare the agrarian roots of industrial growth showing, for example, how Africa differs from Southeast Asia. These authors, however, tend to attribute differences to policy rather than the basic structures within which policy is made and implemented.

This chapter takes a bolder yet necessary approach that builds on the insights of early Japanese Africanists and their colleagues. It first locates Africa in the history of humankind and then explores the distinctive non-agrarian features that characterize contemporary rural Africa. In so doing, the chapter frames the content of the case studies that follow.

The Imanishi legacy at Kyoto University

Our thesis about non-agrarian societies builds to a great extent on the intellectual tradition of the Kyoto School of Field Science which originated with Professor Kinji Imanishi (1902–92), a renowned biologist-alpinist.[1] From the 1930s, Imanishi organized scientific expeditions overseas, in which he trained a number of field scientists from Kyoto University including biologist-turned-ethnologist Tadao Umesao (1920–2010) and ethnobotanist Sasuke Nakao (1916–93). In the postwar period, Imanishi launched a study group on primates together with Jun'ichiro Itani (1926–2001) with the intention of exploring the origins of the human family. Their research field expanded from Japan to Africa and other parts of the world. While their research developed into the study of great and lesser apes, they also started fieldwork on hunter-gatherers in Africa. The research on primates and humans was thus coexistent at the beginning under the same Big History theme of exploring the origins of human society.

In 1959, Imanishi became a professor at the Institute for Research in Humanities at Kyoto University and launched a study group on the evolution of humankind. Along with Umesao, Nakao, and Itani, a philo sopher, Shumpei Ueyama (1921–2012) joined the group. Partly based on the discussion raised in the group, Umesao proposed an ecological interpretation of the history of civilizations (Umesao 2003), which had a great impact on postwar arguments about Japanese society.[2] Ueyama, for his part, struggled to establish a more universal model of human history, in contrast to the Marxian theory of historical development, which was, in his view, full of western bias. Unlike the dialectical yet unilinear model of historical development proposed by the Marxists, Umesao and Ueyama demonstrate that human history involves multiple paths based on the ecological conditions of each broad region of the Old World. These two theses together provided their colleagues, especially field scientists in Kyoto, with a common framework of analyzing and understanding human history. Neither of these theses focused specifically on Africa but the framework lends itself to highlighting the dramatic difference between the historical paths taken by sub-Saharan Africa and other regions.

Ueyama's three-stage theory of human history

In his insightful essay on the development of social formation, Ueyama (1966) divides the history of human civilization into three broad stages: Natural Society, Agrarian Society, and Industrial Society. Table 2.1 below shows social formations in these stages along with predominant modes of exchange at each stage.

According to Ueyama, Natural Society is a stage in which human livelihood is essentially determined by the natural environment. Basic social groups in Natural Society are families and kindred organizations such as the clan. A full-fledged state or stratified society is absent at this stage. The Agrarian Revolution[3] is important as marking the transition from Natural to Agrarian Society, just as the Industrial Revolution marks the transition from Agrarian to Industrial Society.

The Agrarian Revolution is not a synonym for Agricultural Revolution (Braidwood 1960), which means the domestication of wild plants and animals, or its close synonym, the Neolithic Revolution (Childe 1936). Ueyama used Agrarian Revolution to denote a series of revolutionary events including mass production of grain based on irrigation systems, and the creation of urban centers, both of which were related to the emergence of the ruling class and state institution. In this sense, Ueyama's concept of the Agrarian Revolution embraces both the Neolithic and Urban Revolutions proposed by Childe (ibid.). In the long course of history, the Agrarian Revolution gave birth to centralized state power based on institutions such as bureaucracy, the legal system, letters, a

Table 2.1 The historical development of social formations

Units of Society	*Natural Society*		*Agrarian Society*		*Industrial Society*
Family* (Household)	Family 2	*Agrarian Revolution*	Family 3	*Industrial Revolution (modern agricultural revolutoin)*	Family 4
Kindred or local community	Community 1		Community 2		Community 3
State	Non-existent		State 1		State 2
Predominant mode of exchange**	Reciprocal exchange		Plundering and redistribution		Market exchange
Agriculture**	Agriculture 1		Agriculture 2		Agriculture 3

Sources: Adopted from Ueyama (1962: 163) and Karatani (2014).

Notes

* Ueyama (1962, 1966) assumed that there was a pre-Natural Society stage, in which societies of humankind were made up of a group of families or oikia (Family 1), rather than assuming the form of a two-tier organization made up of families and consanguineous communities.

** The rows on the mode of exchange and agriculture are the author's own addition to the original diagram made by Ueyama.

standing army, and religion; and people were mobilized by the state for developing agricultural production e.g., public works to construct irrigation facilities.

The Agrarian Revolution, characterized by large-scale irrigation agriculture and mass-production of cereals, first took place around 4000 b.c., leading to the rise of ancient civilizations like that in Mesopotamia. This landmark development in the production system was accompanied by the emergence of a multilayered social structure, which eventually led to the emergence of states. Civilizations whose agriculture was dependent upon large-scale waterworks for irrigation and flood control were called "hydraulic civilizations" by Karl Wittfogel, the German American historian, in his seminal work *Oriental Despotism* (Wittfogel 1957). This "agro-managerial and agro-bureaucratic" characterization is not necessarily associated with irrigation facilities alone, but with the presence of large-scale public works in general.

The Agrarian Revolution eventually paved the way for the creation of Agrarian Societies not only in the Middle East and the Mediterranean regions, but also in other drylands such as northern China and western India by 2000 b.c. The agrarian civilization gradually advanced into forest lands including northern Europe, Southeast Asia, and Japan, where the creation of states can be observed in the early or mid-first millennium a.d. From the late eighteenth century to the late nineteenth century, the Industrial Revolution took place first in Britain and other parts of Europe, and then in Japan. By this point in time, some advanced nations had finally attained the stage of Industrial Society.

This schematic periodization was proposed by Ueyama as an alternative to the Marxian view of historical development from Asiatic to feudal and then to capitalist modes of production. Within his own framework, Ueyama assumed multiple paths of development. Feudalism, which matured in Western Europe and Japan, is only one form of Agrarian Society. Another important point about Ueyama's scheme, as shown in Table 2.1, is that the progress of civilization never eliminates the basic units of social formation, such as families, kindred, and local communities. They have continued to exist throughout historic times, though always changing in their forms and relations to upper social structure. Likewise, modes of subsistence such as farming, hunting, gathering, and pastoralism have also adapted according to the shift from one stage to another. For instance, pre-modern agriculture (agriculture 2 in Table 2.1) in Agrarian Societies was transformed into modernized agriculture in Industrial Societies (agriculture 3 in Table 2.1).

The transition from Natural to Agrarian Societies

Most parts of the world made a transition from Natural to Agrarian Society in the early stages of history. Whether feudalism or despotism, this social formation was characterized by the effective state control of peasant production and was based on stratified and sedentary farming communities. Peasants existed as a subordinate class, and their farming system was characterized by grain-centered agriculture with staple crops being unified and standardized. Creating new environments for agriculture, typically irrigation systems, was one of the major hallmarks of Agrarian Societies, as Wittfogel demonstrates. Unlike Natural Society, Agrarian Society has a fully developed state system – we call it "agrarian state" system – based on the control of peasant communities and their agricultural production. Africa has had its share of state systems – kingdoms and empires – but the question is whether they qualify to be called "agrarian states."

In his book on state making in ancient Mesopotamia, Scott (2017) provides several clues to the origin of agrarian states as well as their relations to surrounding non-state people made up of mobile hunters and gatherers, pastoralists, and shifting cultivators. Like Ueyama, he pays special attention to the emergence of grain cultivation on a large-scale, which eventually paved the way to state formation. Early states, not only in Mesopotamia but also in Egypt and China, were all dependent on the concentrations of grain production and manpower. Scott also insists that early states were often short-lived and unstable, as it was difficult to tame and subordinate (or to "capture") people who were originally freer and could depend on mobile and dispersed subsistence strategies rather than the mono-cropping associated with fixed-field cultivation. Although he does not mention Africa nor use the term Natural Society, there is a striking resemblance between his description of non-state people and rural dwellers in precolonial (and postcolonial) Africa.

The demarcation between state and pre-state societies is also at the core of the theory of another Japanese philosopher, Kojin Karatani, who analyzed the structure of world history from the viewpoint of modes of exchange rather than modes of production (Karatani 2014). He argues that pre-state polities like the chiefdom (as the extended form of a clan society) are characterized by reciprocal exchange, while in a state society, plunder and redistribution (by a centralized political power) is the dominant form of exchange.[4] The economic principle of the agrarian states is plunder and redistribution of peasant surplus products by a king or a feudal lord. Africa's precolonial "states" should be re-examined in terms of which mode of exchange, not production, was predominant in particular polities.

An especially interesting point in Karatani's analysis is that he regards clan society as an ingenious invention to prevent inequality and the eventual formation of a state. Unlike the despotic ruler of a state, the tribal chief cannot hold absolute power. His position is secure only as far as he continues to make gifts of his wealth to others, and the ever-emerging gap in wealth and power is mitigated. A higher community or alliance of tribal or clan groups may be formed, but the subgroups retain a definite degree of independence. Thus, a clan society (or chiefdom) is not a society that has failed to form a state structure; rather, it is a system that has been able to avoid state formation (Karatani 2014: 9). Another important point made by Karatani is that agrarian communities were created by the state through the process of reorganizing existing clan-based communities.

> For example, in Sumer the state mobilized increasingly large numbers of people to engage in large-scale irrigation projects, granting them land in return for their work. It was thus the state that created the agrarian community. Other than requiring payment of tribute and forced labor, the despotic state did not interfere in the internal affairs of the agrarian community, in which regular systems of self-government and mutual aid existed. At first glance, these may appear to be remnants of the principle of reciprocity found in clan society. But here the autonomy with regard to higher-order organizations that characterized clan societies no longer exists.
>
> (Ibid.: 74)

Thus, both Scott and Karatani criticize the dogma that human beings are inherently in favor of a sedentary mode of life and of being a state subject. Although from different perspectives, both stress the difficulty of "capturing" non-state people who have their own alternative subsistence activities and modes of exchange. Both these theses are relevant for analyzing the formation and dissolution of polities in Africa. At the same time, we must ask what made the ephemeral and unstable early states and civilizations in other regions than Africa so enduring and permanent, lasting for millennia until the present. To answer that question, it is not enough to know the history of each individual state but rather place them in a comparative civilization perspective.

In many societies, the transition from a Natural Society to an Agrarian Society took place by introducing cultures and institutions from more advanced cores of civilization. For example, Japan, like sub-Saharan Africa, is located on the periphery of the Great Civilizations of the Old World. Initially regarded as uncivilized "barbarians" by neighboring Imperial China, Japan transformed itself by introducing "devices and institutions" as technologies of governance from the Chinese World, including its writing system, weights and measures, laws, urban planning, political and religious ideologies, and the institution of bureaucracy (Umesao 1984, 2003). Ueyama (1990) argues that the Agrarian Revolution in Japan was completed around a.d. 700, when a system of centralized government based on the *ritsuryô* codes (codes of laws and ethics originally imported from China) was established. The secular legal system had to be supplemented by a universal religion in order to govern people from within. The then Japanese Emperor Shomu (enthroned in 724 as divine king) tried to create an imitation of the Chinese Empire based on Buddhist principles.

Before that, Japan had been plagued by endless violence between clans, which was finally overcome by the adoption of the universal principles of Buddhism (Umesao 1989: 252–5). The important point is that there is an unbroken historical continuity from the eighth century *ritsuryô* state to contemporary Japan. Japanese civilization has during this period developed without any significant interruption. Such a cumulative development of civilization never occurred in most parts of Africa south of the Sahara.

Precolonial sub-Saharan Africa as Natural Society

Africa is the exception to the general pattern of historical development in the Old World. The sub-region has never developed long-lasting centralized agrarian states nor a persistent subordination of the peasants, the only exception being northern Ethiopia. Communities in sub-Saharan Africa remained Natural Societies. In modern times, Natural Society has persisted in various parts of the world, not only in Africa but also in Polynesia and other numerous outlying areas. The African continent, however, is exceptional in the continued massive presence of Natural or pre-Agrarian Societies.

The concept of Natural Society resembles "primitive society," a concept which has recently been criticized from the viewpoint of political correctness (Hallpike 2011). "Primitive society" is indeed a relative and ambiguous concept which has been used to denote backward and simple livelihoods when compared with civilized societies. Natural Society is a more refined concept as a historical framework based on a clear stage theory. It includes both so-called "primitive" societies and clan societies[5] and is defined as a society which has not experienced an Agrarian Revolution.

Natural Society is also different from the notion of "natural economy" which has often been employed in western historical studies and Marxist literature (e.g., Luxemburg 2003). It tends to be used as a concept that contrasts with cash economy (or capitalist economy), carrying the overtone of a self-sufficient economy without commodity production. Henry Bernstein (1979) refers to "natural economy" (a social formation in which the production of use-value is dominant) as a historical baseline of Africa. The concept, however, is not helpful for distinguishing between Africa and other societies as natural economy can also be found in Agrarian Societies.

Natural Society is the pre-agrarian stage of human history, but pre-agrarian does not mean "pre-agricultural." In fact, many communities in Natural Society were agricultural communities. Agriculture existed both before and after the Agrarian Revolution; however, it went through an essential change in the course of transition from Natural to Agrarian Society. It is the same as what happened in the transition to Industrial

Society: a manufacturing industry existed both before and after the Industrial Revolution, yet its characteristics changed considerably and reconfigured to be aligned with a capitalist economy after the revolution.

Africa's unique path as non-agrarian society

It is important to distinguish between the origin of agriculture and the origin of Agrarian Societies. According to noted Japanese ethno-botanist Sasuke Nakao (1966), Africa is home to one of the oldest forms of agriculture and important domestic plants including millets, sorghum, sesame, and cow peas. He argues that the "Savanna Agriculture Complex," one of the four great agricultural systems of the world, may have originated in sub-Saharan Africa. This might have been as old as another great agriculture system, the "Mediterranean Agriculture Complex," which eventually gave birth to the great ancient civilizations of Mesopotamia and Egypt. When the latter was at its zenith, Mediterranean agriculture accepted cultivars that had been developed in the Savanna Agriculture Complex in sub-Saharan Africa. The latter, however, did not accept any notable crops or innovation from Egypt except for domestic animals. Those practicing the Savanna Agriculture system may have believed it was so perfect that they did not have to accept other crops from outside.

Thus, in the earlier stage of development of agriculture, sub-Saharan Africa might still have been superior or equal to other regions in terms of farming technique. The subcontinent not only deserves respect as the birthplace of humankind, but also as the cradle of one of the most important agriculture systems and crops. Its place in human civilization, however, has since largely vanished and the question is why.

Limited influence of advanced civilizations

Sub-Saharan Africa's geographical position in relation to great ancient civilizations as well as its natural environment are decisive factors in shaping its history. It is closer to the Mediterranean-Islamic World[6] than to any other centers of ancient civilization. Some of the early polities developed under the influence of economic and religious ideas borrowed from Islamic (or Christian) civilizations. Only northern Ethiopia, as a periphery of Mediterranean-Islamic Civilization, developed a polity close to an agrarian state with a formal bureaucracy, local aristocracy, and heavy taxation of peasants who used ox-plough to cultivate grain.[7] There is an apparent historical continuity between ancient Axum and the present-day Ethiopian civilization. Except for this Ethiopian case, however, most polities (mainly established under the impact of Islam) did not nurture a lasting state structure. Early Islamic influences were later largely (but not wholly) replaced by those of Western Europe, which had a greater and farther-reaching impact on the continent's socioeconomic structure.

Elements of external civilizations encroached very slowly upon African societies. In attempting to compare civilizations in Africa and Southeast Asia, Kakeya (2018: 206–10) divides the sub-Saharan region into three major parts: Sahel, Swahili, and Inland Africa. Located in the southern periphery of the Sahara, Sahel has been deeply involved in trans-Saharan trade with the Arab world. The Swahili coastal area in East Africa is located at the fringe of the wider Indian Ocean trading networks. Sahel and Swahili, as culturally coherent areas, were formed through interactions with the outside world as a periphery of the Mediterranean-Islamic World. Inland Africa, on the other hand, remained relatively isolated from such external influences.

On the East African coast, over the centuries, with a constant migration of Arabs and Persians, the Islamic urban way of life has characterized Swahili culture, which flourished in port towns from the thirteenth century or even earlier. In contrast to Sahel, however, the influence of the Swahili civilization was largely limited to a narrow strip of the coast, probably because of the unfavorable, drier environment of the hinterland (Connah 1987: 176). Swahili political organizations largely remained as city-states based on trade and had only limited impact on inland agricultural communities.

It is likely that only Sahel (along with forest and coastal areas south of the Sahel) could have produced the full-fledged states that would match those of other Agrarian Societies. Sahel, a semi-arid area south of the Sahara, saw the rise and fall of "empires" with a wide territory. In Sahel, along with its southern forest-savanna edges, numerous polities emerged from the eighth century until the time immediately before European colonization. The Sahelian civilizations, however, were repeatedly disrupted by external forces, preventing a continuous and cumulative historical development. The Ghana Empire (*c.* eighth to eleventh centuries) was conquered by the Almoravids (a faction of Berber nomads) emanating from the north. It was succeeded by Mali (*c.* thirteenth to sixteenth centuries) and Songhai (*c.* fifteenth to sixteenth centuries) Empires, both of which like their predecessor flourished thanks to trans-Saharan trade. Although historical materials are scant, it seems that the Songhai Empire promoted grain production along the Niger River by making waterways and settling slaves captured in warfare. However, the empire eventually collapsed due to the invasion of Moroccan forces (Niane 1984). The Songhai Empire may have subjugated agricultural communities in some spots, but such a "(quasi) state-subordinate peasant" relationship certainly did not spread in a wider area nor did it last long. A tradition of agriculture, language, family system at the grassroots may have continued to exist but the imperial political institutions did not. Unlike other Agrarian Societies, these pre-modern African polities have had virtually nothing to do with the framework of state formation in modern Africa. The Atlantic slave trade initiated by Europeans and the eventual colonization put an end to any endogenous development of local civilizations.

Non-agrarian features of African polities

In many cases, the power of the African polities rested on the control of trade rather than agricultural production. With a sparse population scattered across open savanna or tropical forest, it was easier for pre-modern African polities to tax long-distance as well as local trade (Connah 1987; Iliffe 1995). By examining many precolonial polities in Africa, Coquery-Vidrovitch (1978) concludes that traditional African societies had a dual social structure: power and elite based on long-distance trade on the one hand, and subsistence and autonomous village economy on the other. She asserts that the revenue of precolonial polities mostly came from either long-distance trade or warfare (in search of booty) rather than from levies collected from villagers. Rural Africa in precolonial times was also unique in the absence of land alienation by an upper class. To quote, "in Black Africa, the sovereign exploited the neighboring peoples, not his own subjects" (Coquery-Vidrovitch 1988: 57).

The economy of Sahelian Empires and kingdoms was based on rule over scattered "desert ports of trade." Rulers were mainly concerned with the control of mines and trade routes rather than increasing agricultural productivity (Shimada 2006). It is not surprising, therefore, that there is little evidence of intensive agriculture in precolonial West Africa; at best irrigation based on natural flooding as practiced in the inland Niger Delta and Lake Chad (Connah 1987: 111; Shimada 2006), or the use of manure. Given the importance of trade and warfare, raising livestock (camels, donkeys, and cattle as pack animals and horses for cavalry) appeared a higher priority than an increase of agricultural production. After his field trip to the Sahel, Sasuke Nakao (1969: 53, 102) made the following insightful remark on what he labeled "anti-agrarian states":

> (Referring to Hausa who had started to use ox plough under the pressure of increasing population) the most advanced area in Africa is now entering the phase of (ancient) India and China. Sahelian Savanna agriculture's origin is as old as wheat cultivation in West Asia, but the former lagged several thousand years behind the latter, with villages scattered throughout a boundless expanse of the savanna, and without utilizing domestic animals fully because of the threat from wild animals and tsetse flies.
>
> (Ibid.: 53)

Escaping state formation

Inland Africa may afford a clue to exploring original African social formations prior to the advent of Islamic influences, because, unlike the Sahel and Swahili regions, it is marked by relatively spontaneous development. Based on Kopytoff's (1987) argument about the African frontier, Kakeya (2018) characterizes Inland Africa as an "internal frontier," in contrast to forestlands in Southeast Asia as an "external frontier" open to maritime trade networks. People living in Inland Africa engaged in frequent settlement relocation and what Kakeya (2018) terms an "extensive mode of life" in contrast to "intensive" production systems which originated in Agrarian Societies. With low population density, Africans utilized their surrounding environment extensively by combining shifting cultivation, hunting (or fishing), and gathering. The village economy was largely self-sufficient, and the salient feature of village society was the leveling mechanism based on the norm of reciprocity. The basic type of social formation was a loose federation of small chiefdoms. In some cases, these chiefdoms developed into "kingdoms" with a wider territory (often based on long-distance trade), but each village's subsistence economy was retained.

Kakeya and Kopytoff demonstrate the highly fluid nature of group formation in Inland Africa, characterized by the constant ebb and flow of polities. While there were repeated attempts to form a state-like structure, there was also constant movement of residents to escape from state formation. James Scott's (2009) argument about "escape agriculture" is illuminating in its capacity to help us understand the movement of the inhabitants in Inland Africa. Analyzing "stateless" hill tribes in mountainous areas of Southeast Asia, Scott argues that these people deliberately selected certain farming systems and crops to "escape" from the mighty arms of the lowland states based on wet rice cultivation. Shifting cultivation and adoption of "escape crops" such as sweet potato, cassava, and maize were particularly important strategies employed to escape from control by lowland agrarian states.

Bypassing the agrarian stage

The slow European penetration of Africa during the nineteenth century began to escalate into a scramble for territory during the late 1870s. Natural Societies in rural Africa were abruptly assaulted by modernization projects initiated by the imperial western powers. Unlike what happened in other regions, in the twentieth century the peasantry in Africa experienced a direct transition from Natural to Industrial Society without passing through the stage of Agrarian Society. Because this transition is still incomplete and has faced many unforeseen obstacles, social formations in Africa are not the same as in those countries which evolved by passing through an agrarian transformation.

As for modernization of agriculture and rural life, what colonialists and postcolonial governments found in rural Africa were elusive, dispersed, and self-sufficient communities of cultivators, pastoralists, agro-pastoralists, and hunters and gatherers, who frequently moved from one place to another. The process of incorporating these people into a modern system of state and economy was not straightforward but full of twists and turns.

There have continued to be rivalling impetuses to engage or exit market and state. It is for this reason that African peasants in the twentieth century appear sometimes to be capitalistic, and at other times to be averse to the market. The coexistence of highly commercialized farmers and subsistence peasants is one of the hallmarks of contemporary African villages. This is in sharp contrast with villages in Southeast Asia where the majority of the farmers rushed to embrace agricultural commercialization at the expense of existing socioeconomic systems, as further discussed in Chapter 7.

Non-agrarian features of contemporary rural Africa

The analysis of African livelihood systems in mainstream western ethnography tends to focus on the differences that scholars see, for example, between hunter-gatherers, pastoralists, and sedentary cultivators. In the Big History perspective of the Kyoto researchers, not only is it that the commonalities which exist among these communities are seen but also how they have influenced each other that matters. Thus, they highlight the legacy of Natural Society imprints in contemporary rural communities in Africa with a special emphasis on how non-agrarian features shape agriculture.

The Itani School's perspective on rural communities

Jun'ichiro Itani who began his research by studying primates later moved into detailed studies of the ecological management of Africans living with nature. Itani and his disciples initially focused on the societies of hunter-gatherers like the San in the Kalahari Desert and the Pygmies in the rainforest of the Congo Basin who were considered to be preserving the most primordial form of human living. Some of them later moved their interest to studies of cultivators and pastoralists. Their main common theme was to pursue human-nature interaction among people who are deeply dependent on their natural environment ("naturalists" as Itani called them). High mobility and dispersion make these naturalist societies in Africa unique. Another theme that was frequently debated among the Itani School researchers is what they saw as the egalitarian nature of these communities, displayed notably in the distribution of foods. The quantitative and qualitative analysis of this "economy of sharing" was conducted not only in communities of hunters and gatherers but also among cultivators and pastoralists (Kakeya 1976; Sato Hiroaki 1983; Itani 1988; Sato Shun 1992; Kitanishi 2000; Ichikawa 2005).

Although it is commonly accepted that hunter-gatherers and cultivators are ranked at different levels of human evolution, the above studies indicate that there are striking similarities between cultivators and hunter-gatherers in Africa, especially in terms of their flexible social organization, high mobility, and economy of sharing. These are features of Natural Society which have survived major political and economic changes to this day. Peasant cultivators, hunter-gatherers, and pastoralists share these elements in the sense that their communities are bound together by sharing consumptive goods rather than immovable productive resources. This is in contrast with agrarian communities where villagers are tied to a specific piece of land and incorporated into a stratified social structure with a fixed group membership. As we shall see below, this suggests that agricultural communities in Africa are closer to hunting-and-gathering communities in the same continent than they are to farming communities in Agrarian Societies.[8]

Thus, based on meticulous fieldwork in rural Africa, the Itani School of ecological anthropology has filled the gaps left in Ueyama's argument on Natural Society. The latter only noticed that a Natural Society is a single-stratum society as opposed to a stratified Agrarian Society. The Itani School of Africanists have gone further by empirically demonstrating that sharing and redistribution of wealth to prevent social stratification is a hallmark of Africa as a Natural Society. Their rich empirical studies have greatly contributed to our understanding of the close interaction between man and nature.

Among the early Africanists in Kyoto, the late Professor Makoto Kakeya (1945–2013) was especially influential. He carried out detailed field research among the Tongwe on the eastern shores of Lake Tanganyika and the Bemba in northern Zambia, both of which are ethnic groups living in the same type of savanna woodlands (Kakeya 1976, 1986; Kakeya and Sugiyama 1987; Kakeya et al. 2006). As ecological anthropologist, his primary intention was to examine the physical and biological base of subsistence among shifting cultivators, but he went further by also delving deeply into the spiritual and supernatural world of the villagers (Kakeya 1982). In his fascinating study of the Tongwe, Kakeya demonstrates how despite inequality in crop production, food consumption of each household is eventually evened out through the active sharing of foods. A similar "leveling mechanism" was also observed by him and his colleague Yuko Sugiyama among the Bemba villagers in northern Zambia.

Kakeya provided unique insights into rural Africa by focusing on the "mode of life" of the shifting cultivators showing how their system of production and consumption, along with their value system, is extensive and thus different from agrarian communities based on intensive farming. In Table 2.2 Kakeya identifies ideal types of "the extensive mode of life" (of African shifting cultivators) and "the intensive mode of life" (of modernized farmers, e.g., in Europe or Japan). The latter developed specifically in the middle-latitude forest areas of the Temperate Zone (Japan and Western Europe), where the mentality of maximization of production based on private ownership is predominant and the ensuing socioeconomic inequality is ubiquitous. The extensive mode of life, found typically among African shifting cultivators, is characterized by modest production within the limits of the natural environment. Using this framework,

Table 2.2 Two modes of life

Extensive mode of life	*Intensive mode of life*
Extensive farming	Intensive farming
Farming based on low population density	Farming based on high population density
Labor-productive farming	Land-productive farming
Polyculture	Monoculture
Mobile	Sedentary
Common ownership	Private ownership
Generalist of utilizing natural environment (less specialized in farming)	Specialist of utilizing natural environment (specialized in farming)
Stability	Expansion
Minimum subsistence effort (Underproduction)	Maximum livelihood effort (Overproduction)
Levelling mechanism	Social differentiation
Centrifugal	Centripetal
Segmentary (political system)	Centralized (political system)

Source: Kakeya 2018: 286.

Kakeya tried to elucidate conditions which have supported the stable reproduction of communities and nature among traditional shifting cultivators. Although he seldom used the term Natural Society, we show in this volume that most of the characteristics of an extensive mode of life can also be found in other non-agriculturist communities that are still shaped by a Natural Society logic. As subsequent empirical chapters in this volume demonstrate from a variety of ecological and economic settings, peasant cultivators in Africa have more in common with their hunting and pastoralist neighbors than they have with farmers in Agrarian Societies. This is an underlying structural factor which explains why an intensive farming approach associated with modernization does not fit into existing, indigenous extensive modes of life.

Conclusions

There has been a persistent "agrarian bias"[9] in international development policy circles which is why African governments have often been blamed for not implementing policies of agricultural intensification. As this volume will discuss further, such blame is misplaced because the recipe itself is faulty. Rural society and farming practice south of the Sahara is not "agrarian" and thus does not lend itself to state-managed policies aimed at intensifying agricultural production. This failure to understand and recognize the peculiarities of rural livelihoods – or modes of life – is a major reason why agricultural productivity on the land in Africa has continued to fall behind the rest of the world. According to Hirano (2013), grain productivity in sub-Saharan Africa remains on average one-third of the world average (one-third in maize, half in rice, less than one-third in wheat). That is why the argument and findings presented in this volume need to be considered more closely not only by the research community, but also the policy world where the agrarian bias is most deeply rooted. Agricultural development in Africa will never fully materialize until it is built on its own different premises. The persistence of a Natural Society legacy is not a practical policy problem but rather an intellectual nut that waits to be cracked. The rest of this volume will provide some of the tools needed for that task.

Notes

1 One of Imanishi's major works was translated into English (Imanishi 2002).

2 As for the influence of Ueyama's and Umesao's argument on evaluation of Japanese society, see Morris-Suzuki (1998: 143).

3 In the original Japanese text, this term was referred to as *nôgyô kakumei*, which was translated from agricultural revolution. The agrarian revolution in the context of European history often refers to a change of social structure of agriculture in modern times in relation to the changes in ownership of land (from the landed gentry to peasant). It may also refer to land reform in contemporary developing countries (Paige 1975). We prefer to use the term agrarian revolution in order to distinguish it from the agricultural revolution as a close synonym to neolithic revolution, and partly because it demarcates the transition to agrarian society.

4 It is not to say that, in pre-state societies, reciprocity was the sole mode of exchange. There was trade, as well as plunder and redistribution to some extent even in pre-state societies. Likewise, in an Agrarian Society, while plunder and distribution is the dominant mode of exchange, trade is active in towns, and reciprocity persists in local agricultural communities. The point is which mode of exchange takes the leading role (Karatani 2014: 9).

5 Karatani (2014) distinguishes primitive societies based on "pooling" as a dominant mode of exchange and clan societies (including chiefdom and tribal kingdom) based on sedentary communities and reciprocity.

6 The Mediterranean-Islamic World is a term used by Umesao (2003). It denotes a series of civilizations that rose and fell in Mediterranean areas including Roman, Byzantine, and Islamic Empires.

7 It is worthy of note that Ethiopia was the only precolonial society in sub-Saharan Africa which had a significant tradition of map making and plough cultivation, both of which are related to the effective control and appropriation of land (Herbst 2014: 38–40).

8 In a special issue of *Current Anthropology* ("Intergenerational Wealth Transmission and Inequality in Premodern Societies," vol. 51, no. 1, 2010), there was a debate on egalitarianism and wealth transference (from one generation to the next) in rural communities, where scholars came to a similar conclusion about the similarities between hunter-gatherers and subsistence peasants. On the other hand, societies based on intensive agriculture tend to have an economic gap deriving from the private ownership of land, which can be inherited and accumulated from one generation to the next.

9 The term agrarian bias has so far been used by development experts to mean the preference for the agricultural sector over the development of the industrial sector (e.g., Hodge 2007: 262–72).

3 Economy of affection in Africa

The informal basis of development

Goran Hyden

Introduction

At roughly the same time that the Kyoto Africanists conducted their research to uncover the special features of Africa's social formations by relying on a Big History perspective, Goran Hyden was engaged in searching for explanations why the state in Africa was unable to realize its development policy ambitions. Why did the state seem to sit like a balloon over society? Why were African peasant farmers not captured like they had been in other regions of the world? That Africans were no different from peasants elsewhere was the standard assumption among western researchers in the 1970s and 1980s. Whether they were neoliberal or neo-Marxist in their outlook they could not accept the point that the African case would be historically different. The former referred to the power of the market forces and the inability of poor farmers to mitigate their influence around the world (Kasfir 1986). The latter added the argument that colonization had turned African society upside down; the region's precolonial or precapitalist features had been effectively wiped out during the two generations or so that the European colonial powers had administered the continent (Cliffe 1987).

The position taken by mainstream researchers appeared not only imperious but also out of line with more careful empirical studies on how African society operates and politics is conducted that had been produced earlier by anthropologists like Jack Goody (1971), sociologists like Peter Ekeh (1975) and political scientists like Aristide Zolberg (1967). Their works stand out as being genuinely devoted to an examination of the essence of how people live, organize themselves to get things done, and pursue a form of politics that is grounded in the continent's social and economic realities. They provided inspiration for the research that produced the concepts of an "uncaptured peasantry" and an "economy of affection" (Hyden 1980). The two concepts are very much in tune with the research tradition of the Kyoto School although that intellectual connection was established only in the last 20 years when the collaboration between Hyden and the Japanese Africanists began in earnest (see e.g., Kimambo et al. 2008).

The purpose of this chapter is to revisit Hyden's conceptualization and further discuss it in the context of this volume drawing on the arguments put forward in Chapter 2. It will trace the roots of an economy of affection in African modes of livelihood, demonstrate its continued relevance in the absence of corporatist structures, and compare it to other forms of moral economy.

The economy of affection

The economy of affection is an example of the principles of sharing food and other consumables that are associated with Natural Society. Its roots are in rural society but with migration to urban centers and beyond, it is more generally present in contemporary Africa. Members of these groups or communities typically share a common stock of meanings and assumptions to generate and sustain trust and mutual understanding. Reciprocity is the core principle of the economy of affection. Although many Africans, especially in the diaspora, would like to escape the "burden" of sharing, as they see it, the social expectations of sharing wealth to support those who are less well-endowed continue to be influential in societies like those in Africa where the state is not an effective agent of organizing society or facilitating wealth distribution. The economy of affection, therefore, fills an important gap and permeates society in ways that official policy goals rarely do. In fact, its informal institutions serve multiple purposes stretching from survival, subsistence and social maintenance to investment and development (Hyden 1983). This way they tend to be especially important in determining development outcomes.

Other than by anthropologists, little attention has been paid to this set of micro-institutions mainly because in other parts of the world, they tend to be regarded as part of the private sphere or totally dominated by civic and public values. The situation in Africa, however, is different. The logic of Natural Society and the morality of the economy of affection are influential parts of public life. The boundaries that westerners draw between the private and public spheres are hollow in African societies. Formal market or state institutions are modified by informal behaviors and choices with roots in the economy of affection. What most western analysts conceive as "deviant behavior" is in fact often a deeply rooted articulation of how the world is viewed and livelihoods pursued in these societies. There is reason, therefore, to lift this micro-institutional realm to its rightful place in the analysis of political and developmental issues in Africa.

It begins by acknowledging the continued legacy of Natural Society discussed in Chapter 2. It entails not only a material but also a mental or cultural dimension. The colonial powers did their best to overturn both, but they never really succeeded in their ambition. Thus, two decades after countries had become politically independent, one observer described the situation in West Africa in the following terms:

> In the case of West African commercial agriculture, the dominant form of labor is not that done for wages, but that done by a family worker farming on his own account. This farmer is not free from complex social ties: He may be embedded in associations derived from membership in a descent group, and he is most certainly working alongside his wife and children in a setting that is domestic. He may be a sharecropper, splitting his product with a landlord who does not "own" the land and who may be a senior kinsman. He may be working on a settlement regulated by the government or a religious order. All of these are variations of peasant farming and they still predominate.
>
> (Hart 1982: 117–18)

Hart's characterization confirms the persistence of social formations that have not been subordinated to the commands of an agrarian state. Local variations could be found in other parts of the sub-Saharan region and they persist also there. For example, Goody (1971: 24–5) shows that the origin of these forms is not in differences in mercantile activities. Like in medieval Europe, trade was highly organized in both precolonial East and West Africa. Whether societies were kingdoms or stateless, barter had given way to more complex forms of exchange. The origin instead can be found in technology (or perhaps better said – lack of it). Unlike Eurasian societies in the Bronze Age, Africa never adopted the plough which made possible a substantial rise in productivity that in turn meant a greater surplus for the maintenance of special crafts, for the growth of differences in wealth and lifestyles, and for expansion into urban-based developments. Furthermore, this process stimulated the move to fixed holdings and away from shifting agriculture with the result that the availability of arable land decreased, and its value increased.

Precolonial Africans made little or no use of mechanical tools, even quite elementary ones. For example, animal power which pulled the plough in Asia and Europe was not applied to agricultural use, a prime reason being that the wheel, despite the existence of an iron-smelting technology, was never adopted for farming purposes. The lack of the wheel also limited the possibilities for water control. To be sure, irrigation existed in precolonial Africa, but it never raised agricultural production to significant surpluses. Hoe farming by men or women prevailed across the region in those days.

It can of course be argued that the reason why there was no agricultural transformation like in Europe and Asia is that there was more than enough land to cultivate for everyone. Land use intensification, therefore, was not a necessity. This had implications for how households and families were structured as Boserup (1970) convincingly argued. In sparsely populated regions where shifting cultivation is used, men do little farm work. Because work falls on the shoulders of women – and often their children – marriage is often polygynous. Furthermore, marriage is a deal where the family of the groom pays bride-wealth to the family of the girl. This transaction differs from what happens in societies where agriculture is more advanced; men do the work, marriages are monogamous, and women tend to be valued only as mothers. Here, the marriage transaction goes the other way: a dowry is paid by the girl's family to that of the bridegroom typically interpreted as a contribution to the livelihood of a nuclear household.

Goody has elaborated on this difference which sets Africa apart from Eurasia by showing that inheritance patterns also vary. Because parental property in Africa is shared in the marriage transaction, wives can make no claim to the husband's property upon his death. Instead, such property is transmitted only to members of his own clan or lineage. Women are not entitled to share in the property of their father's or mother's brothers, even when they are members of the same unilineal descent group (Goody 1976: 6–8). This lateral transfer of property to brothers and sons but not to daughters stands in contrast to Eurasia, where the plough and use of irrigation have allowed for more advanced forms of cultivation. Here the transmission is vertical, i.e., within monogamous families allowing both male and female children to have access to the estate of their deceased parents.

These historical differences between Africa and other regions of the world had consequences for social organization and social differentiation. As we have shown in Chapter 2, there were differences in wealth, but stratification was due not to capital but labor. Chiefs and other potentates owed their power to political office and not to being landlords owning land on which others were tenants. They ruled through and over people, not land. Polygyny was used to cement social bonds across families and households. The Kingdom of Swaziland where almost every family is somehow related to the king is a still live example of social formation based on consumption and social reproduction rather than production. Number of subjects was more significant than territorial size. It is instructive that in Ethiopia where the Amhara had the plough, land took on considerable economic value. The use of marriage transactions to maintain social status, like in Europe and Asia, led to forms of social stratification not found elsewhere on the continent. Ethiopia was African in a geographical but not a social sense.

Elsewhere on the continent, even in centralized kingdoms, social organization was fluid and rarely based on differences in landholding. Land was owned communally by a lineage group though exploited by smaller family units. There was enough land to hold some of it in fallow and no one had to beg on his knees to have access to land. As we showed in Chapter 2, sharing of food was a principle of honor for the already wealthy. For these reasons, African societies did not develop feudal patterns of social organization characterized by serfdom or peonage. Slavery was widespread but yielded only minor payoffs. Clientship existed but the voluntary subjection associated with such transactions involved cattle, as in Rwanda, or political rights, e.g., to collect tribute or taxes. In fact, as discussed in Chapter 2, income from taxing trade and gifts from traders contributed more to the income and status of rulers in precolonial Africa than rent or tribute from agricultural activities ever did (Goody 1976: 104–12). It is worth noting that social and political violence in postcolonial Africa has not really been directed against the upper strata of society. Peasants have risen against indigenous rulers only in Ethiopia, Rwanda and Zanzibar where marriage rules were meant to isolate the ruling caste. Elsewhere, conflicts causing civil violence have typically been between groups organized along communal or religious lines. These facts clearly show that African societies have never produced a coherent "peasant class", which was normally created in other regions through a long history of subjugation of peasant cultivators by an upper class.

The absence of a transformation to Agrarian Society and an intensive form of agriculture that raises the value of land left sub-Saharan Africa with a degree of fluidity in social relations and the rise of an economy in which trade – and the ability to control and tax it – became the prime source of social stratification. Chiefs became rich and influential using their political office but really did not differ significantly from commoners when it came to owning land. To borrow Sara Berry's phrase, that "no condition is permanent" in rural Africa is as applicable to precolonial times as it is to later days (Berry 1993). With the legacies of a Natural Society living on in virtually all spheres of life, status rests on the ability to manage environmental challenges – both agrarian as well as non-agrarian – and a readiness to share food and other consumables with members of self-defined communities that go beyond the confines of a nuclear family. That is why the argument that social formations in Africa are reflections of the presence of a distinct "economy of affection" makes sense.

Why it is still relevant

Its micro institutions continue to be present in post-colonial Africa. The question is why it has lasted into the twenty-first century. This issue needs attention because scholars interested in African politics and development have too easily dismissed its precolonial legacy. Many have assumed that colonialism eradicated this legacy, but as Crawford Young (1994) has argued, in a comparative perspective, the colonial state in Africa never had the same impact as it had in Latin America. First, colonialism in Latin America occurred at an earlier point in time and lasted longer. Like Africans, people in Latin America had to struggle to achieve their political independence, but they did so at a point in history before the values of liberalism had become universalized. Independence meant local aristocrats replacing those loyal to the Crown in Spain. The colonial presence in Africa, in contrast, began only when the first wave of democratization in

Europe and Latin America was just around the corner. And it ended at the crest of the second wave around 1960. It lasted only two generations. For example, Kenya's first president, Jomo Kenyatta, was born before the British had established control over the territory and served as its head of state for 14 years before his death in 1978.

A second reason for the persistence of the economy of affection is that the colonial state was hardly a Hobbesian Leviathan. To be sure, Africans were treated brutally in many places, Belgian Congo being the foremost example of brutality against the local populations. Its ability to govern, however, has been called into question by several analysts, e.g., Herbst (2000), because unlike Europe where states were formed through wars (Tilly 1975), colonial states in Africa were created through negotiations between European powers at a conference in Berlin 1894 and, what is more, African government leaders decided after independence to leave those boundaries intact to avoid warfare. African states, therefore, never developed the governance "bite" that turned it into an effective agent of social transformation and sustained development. This may have been more evident in British than in French colonies because they were more inclined to practice a system of "indirect rule," i.e., relying on already existing indigenous authority structures. Even in those places like the French and Portuguese colonies where direct rule was practiced, however, the reach of the state was incomplete, especially in the countryside where communal structures continued to shape livelihoods and the way local communities were governed. The colonial administration aimed at being an agrarian state, but it never developed into one. It was overthrown long before it had reached its objective of socially transforming the African continent.

Third, at the time of independence, government leaders took deliberate steps to reject colonial institutions and reassert what they interpreted as "African values". This meant many different things. It involved the rejection of policies aimed at modernizing agriculture, the abolition of unpopular taxes that gave the colonial state power over households used to being free from any such obligations, and a questioning of the value of civil service neutrality at a time when commitment to the ideological values of the independence project was called for (Rweyemamu and Hyden 1975). These populist measures were not only directed against what in President Trump's America has been referred to as the "deep state" but also taken with the effect of reducing the influence of the state over society. As the state lost its leverage to carry out its policies in an effective manner, it left peasants' uncaptured and affective networks with room to shape the organization of social relations in society along lines that people could recognize as theirs. Thus, the Africanization of the state institutions that followed independence was not merely a matter of training locals to become effective administrators. It had much more fundamental effects on state-society relations and the way these societies would move forward.

Because African leaders after independence ripped up the state of its rationalist character and instead adjusted it to the prevalence of a Natural Society logic, they reinforced the already existing differences with other regions of the world. There is much less evidence of planned interventions aimed at increasing productivity. Land is not divided into squares or rectangular shapes indicating the notion of boundaries. Instead, they are the result of an open frontier mentality. Because land was never a livelihood hurdle in the past, members of the younger generation would move on to new land whenever the consumption needs of a rising number of household members began to exceed what the existing farmland could produce. This readiness to pack up and migrate elsewhere has continued to be a ubiquitous feature of the African scene. It has made migration more stable and less disruptive than in other parts of the world. As Kopytoff (1987: 22) argues, African "roots" are not in a place but in a kin group. Migration is still a common practice although with land becoming scarcer, its direction has changed from the countryside to the cities and beyond.

The contemporary challenge in Africa is not only that societies never went through an agrarian revolution that significantly increased productivity on the land but also that these societies are still to go through an industrial revolution. In modern Europe, industrial and "modern" agricultural revolutions occurred in close sequence and reinforced the social transformation to what we today call modern society. As land use was mechanized and farms consolidated into contiguous units for more effective production, surplus laborers on the land could move to the cities where opportunities for employment were growing. Where jobs could not be found in nearby cities, there was also the chance to migrate to a foreign country. North America and Australia received large numbers of immigrants from countries in Europe seeking a new and more promising life. Thus, nineteenth-century Europe was blessed by historical circumstances that facilitated the transformation to modern society.

Africa's current predicament is different. With the land frontier closing, the option of moving to occupy another piece of unused land is much less of a possibility. Most households are stuck with the land they already have. It tends to get subdivided among members with right to inheritance. This typically means the opposite to what happened in Europe: opportunities for raising productivity decline! This is a major reason why the younger generation leave the village for life in towns or foreign countries even if that means accepting a marginal existence in a new location. The problem for these migrants is that employment opportunities in factories and other large-scale establishments are few. Instead of adopting a new lifestyle as workers living in apartment buildings together with others of a similar economic background, they end up scratching out a living and falling back on the sharing mentality that they grew up with in the village. The luckier ones may help relatives back home by remitting parts of their earnings, but many instead find themselves in situations where they are forced to fall back on the logic of a generalized reciprocity which, for example, means obtaining a "loan" that does not necessarily have to be repaid.

The point is that the economy of affection is not only a phenomenon found in the countryside but is today as present also in urban areas. Relying on sharing within communities of consumption is an integral part of how people in African countries approach life. It is guided by the trust they have in their micro-institutions and the concomitant lack of loyalty they feel toward the impersonal nature of corporate entities.

The absence of a corporate society

Corporations have emerged as the key mechanisms of development in modern society. Emerging economies, like already developed countries, are no longer market societies in the sense of the dominance of an "invisible hand" or a similarly invisible set of informal institutions. Instead, these economies are shaped by hierarchical organizations made up of teams of specialists with their own career ladders. This type of corporate body dominates in both government and business. Their ability to deliver results in an efficient and effective manner has made them models of how to achieve development. Corporate success relies on being able to take advantage of the expertise housed in each of its subunits and coordinate these inputs into managerial decisions that enhance the chances of positive outcomes. Corporatism creates 'The Organization Man' (Whyte 1958) – a person ready to submit to the role of being a cog in a machine – an organizational revolution that has affected virtually all forms of human activity (Boulding 1953).

Corporate Society may be viewed as the ultimate step of alienation that began in eighteenth-century Britain with the alienation of land and later labor. It is the product of a historical process over many generations. Organization has become the prime social technology on which the contemporary world increasingly depends. It has become, as Weber (1968) predicted over a hundred years ago, indispensable for mass administration. As he added in the same context: the choice is that between bureaucracy and dilettantism!

The rise of corporatism in modern society has met with its own intellectual critiques. The most well-known may be the Marxist thesis that capitalism – and by extension corporatism – inevitably leads to the pauperization of the workers. More relevant here though is the view of Habermas (1987) that the expansion of rationality in all forms of social activity is a threat because it brushes aside all forms of institutions based on the mutual understanding of intentions and the general recognition of obligations. He contrasts this trend with preindustrial society in which technical innovation and organizational improvement was contained within the existing institutional life.

The challenge in African countries tends to be the opposite to the pattern identified in Habermas's writings. Development analysts do not see enough of a rationally structured corporate system to effectively organize the pursuit of economic and social development. To be sure, modern organizations exist, but typically more on paper than in practice. Formal attributes are overshadowed or even undermined by the presence of informally shared beliefs with roots in a Natural rather than Industrial Society logic. This means that once policies are made, they often become subject to concerns other than those that are officially stated. The institutional life of the economy of affection tends to take over. This expresses itself in different ways. Social interaction inside the organization is not primarily oriented toward goal achievement. Instead, as for example a survey of public sector managers in southern Africa has illustrated, their main preoccupation tends to be conflicts over turf and money (Montgomery 1987). Policy issues, according to the same source, came only in fifth place on their list of tasks they had to attend to. The moral code inside the organization takes its lead from practices prevalent in precolonial systems of rule over people. The way that it operates has more to do with exercise of power than with a rational pursuit of public goods. The result is that policy implementation tends to produce outcomes that are out of line with stated goals. Intended public goods get waylaid in favor of a more circumscribed form of distribution. These "club" goods benefit only a section of the population and the process reflects the power structure surrounding the allocation of each specific benefit. The idea that power inside an organization matters is universally accepted but what is unique in the African context is that the values that structure its use reflect society at large rather than those associated with the rational character of modern organization (Kiggundu 1991). The legacy of Natural Society is very much present in the African organizational context.

It is this type of issue that has reinforced the belief among donors that public-sector institutions in African countries need to be reformed. The initial attempts after independence focused on training African civil servants and providing technical assistance in the form of expatriate advisors. An evaluation of these programs (Berg 1993) was largely critical of what had been achieved and recommended a major overhaul. By the time that evaluation was published, the approach had already begun to change from capacity-building to institutional reform. A first such waves took place in the late 1970s and early 1980s and entailed above all a call for decentralization as well as privatization of many government functions. It was followed by a second reform wave driven by donor concerns about the effectiveness of their aid. None of these has really turned things around in ways that indicate government bureaucracies in African countries have become more rational and effective in delivering public goods. Andrews (2013) has conducted the most thorough study of donor-funded institutional reforms in Africa and his findings show that these reforms have not really penetrated the cultural frameworks within which these organizations are embedded. The corporate rationality has been applied with too much emphasis laid on domestic accountability principles in the donor countries rather than on an approach that is capable of building institutions from already existing local premises.

Principles of the corporate system have also entered Africa through multinational corporations with investments on the continent. They play an important role in the economy of most of these countries but their impact on changing perceptions among policymakers or ordinary citizens has been limited. One reason is that they have confined their presence to making profits. They have avoided playing a public role other than through forms of corporate social responsibility, for example by financing community development projects or sponsoring sporting events that give them legitimacy in local eyes. Africans who work for these corporations have been shunned by government leaders in many countries because they work for what they perceive as foreign entities. Others in the same kind of employment have simply been individually successful enough to disregard playing a more public role in their home country. The result is that almost all African countries lack a vibrant domestic middle class with the vision and power to institutionalize new forms of organizational life. This is a social change process that takes time and still waits to be fully harnessed in Africa. It is a much more complex undertaking than suggested in "instant" theories of change linked to donor-funded project interventions or logical frameworks created on the assumption that change is a linear pursuit.

The economy of affection today

The economy of affection survives on the strength of a generalized reciprocal logic. It relies on face-to-face transactions to get things done. Its core principles are that (a) whom you know is more important than what you know; (b) sharing wealth is more important than investment for profit; and (c) a helping hand today generates returns tomorrow. It differs from both capitalism and socialism. Money is not an end in itself; nor is the state needed as a redistributive mechanism. As we show in subsequent chapters, the core principles of the economy of affection coexist with elements of capitalism or socialism. They are not erased by capitalism or state policies but rather tend to embed them in their own logic. It is a largely invisible economy that conscientious policymakers loathe, and economists find hard to incorporate into their statistics. These parallel institutions, therefore, tend to be pushed aside and treated as problems to be solved. As studies of anti-corruption efforts in Africa have shown, corruption is much more than a form of deviant behavior. It is reflective of the way social transactions are carried out in an environment where trust continues to lie in face-to-face reciprocities rather than in the impersonal exchanges provided through the market or bureaucratic agencies (Disch et al. 2009).

The economy of affection is not an expression of irrationality or altruism. It is a practical and rational way of dealing with choice in conditions of uncertainty where mental frames tend to be confined to immediately identifiable places. People engage in the economy of affection for a variety of reasons and they may do so from positions of either strength or weakness. Their engagement may come from facing hardship as well as opportunity. Four of the most common motives are to (a) gain status; (b) seek favors; (c) share a benefit; and (d) provide a common good. These motives often reinforce each other and become a cultural pattern. For example, status and wealth in African society have always depended on the ability to accumulate followers and dependents, be they slaves, hired labor, wives or clients. Such accumulation rather than that of land has been the principal way that these societies have been socially structured. Having a lot of money, as this study will demonstrate further, means nothing unless it is shared with others, a point that others have also made, for example, Barber (1991) with reference to the Yoruba people of southwestern Nigeria, and Parkin (1972) showing how among the Giriama people of the Kenyan coast persons with newly acquired wealth spend lavishly on private ceremonies like funerals and weddings to demonstrate their commitment to and respect for local institutions of kinship that make them worthy of appearing before elders in disputes over land and trees.

The economy of affection, however, is much more than prostrating behavior and conspicuous consumption. It plays a prominent role in electoral contests where it reinforces the expectation placed on candidates to deliver goods here and now. This may encourage vote-buying and the promise of goods that are specific to groups of constituents (Lindberg 2003). Beer and other popular consumables are distributed during campaigns to get votes and many of the promises made are related to the delivery of private goods or those that benefit a certain segment of the electorate.

This parallel economy is also much more than clientelism. The latter may have the effect of empowering people to come out and voice their claim for a share of the public good, something that would probably not happen without such face-to-face interaction over the allocation of benefits. The motive for engaging in affective behavior, however, runs deeper because in societies where climate is unpredictable and technology rudimentary, it makes sense to diversify sources of income. This includes relying on relatives and friends that may have relocated to urban areas. Financial remittances from urban to rural areas have long been a prominent feature of African economies (Caldwell 1969; Sandbrook 1982). This does not mean that members of the rural household have been only at the receiving end. As, for example, Andersson-Djurfeldt (2015) has shown, rural households in Ethiopia made their own contribution by remitting maize to urban-based members of their families. The point about these remittances is that they are not guided by contract but by a moral code of generalized reciprocity, i.e., the act entails an expectation of a quid pro quo – however not within a specific time or in terms of equivalent value.

There is a general recognition among development analysts of the relevance of the thesis that "small is beautiful," the phrase once coined by the British economist, E.F. Schumacher (1973). Community self-help initiatives like *Stokwel* projects in southern Africa to build and manage waterholes for cattle or *harambee* efforts in Kenya to build schools and other development structures are among the most well-known micro-institutions that have served rural Africa well. Another form of face-to-face arrangement that has taken off in rural as well as urban areas are micro-credit groups. This type of rotating credit and savings organization goes under many different local names, for example *esusu* among the Yoruba, *adaski* among the Hausa of northern Nigeria, *djana* among the Fang of Cameroon, and *ndjonu* in Benin. They are mutual aid societies that serve important social security purposes especially at times of sickness, bereavement and other unforeseen crises. They are also helpful when individual households want to build a house, pay school fees or acquire goods for trading.

The social logic of the economy of affection is not unique to Africa but because of its Natural Society legacies it is especially prominent there. Individuals are not necessarily autonomous decision-makers in the sense that market theories in economics assume. The individual in the economy of affection makes decisions with the social consequences for others in mind. Being dependent on the goodwill of these others, be they relatives, friends or patrons, any transaction or exchange is social rather than just economic. Both economic and social exchange theory is based on a fundamental characteristic of social life: much of what we value like goods, services, companionship, approval, status and information can only be obtained from others. People depend on one another for such valued resources, and they provide them to one another through the process of exchange.

Social exchange differs from economic exchange in two important respects. Whereas classical microeconomic theory assumes that there are no long-term relations between the partners of exchange, social exchange theory incorporates the premise that enduring relations between the partners exist. Furthermore, classic microeconomics assumes that actors engage in sets of *independent* transactions that are aggregated into markets. Social exchange theory, in contrast, is built on the assumption that actors engage in recurring *interdependent* transactions. In short, what distinguishes social exchange theory from microeconomics – and modern psychology, for that matter – is its emphasis on social structure as the context in which exchange processes take place. The economy of affection can only be adequately understood through social exchange theory.

Social exchanges are informal, nonnegotiated and hence not driven by adherence to an abstract norm or rule but the dynamics of exchanges involving rewards and punishments in different degrees and at intermittent intervals. As parallel institutions with those formal entities that modernity has produced, they are seen by outsiders as problematic and a direct threat to development. For most Africans, however, these informal institutions are viewed as important mechanisms for securing their livelihoods. When they create their own narratives of what is going on, it is these informal institutions and their underpinning values that feature most prominently. Their stories center on one or more of the following informal institutions:

Clientelism which tends to permeate social and political life because it offers gains for both patrons and their followers. It is generally a preferred way of obtaining a scarce but difficult to obtain good. Being on good terms with somebody with power in society, therefore, is an important consideration that has its own consequences for social stratification.

Pooling which involves cooperative entities that are organized along self-enforcing lines. They are not sanctioned by law but constituted by adherence to often unwritten rules. The credit and savings groups, discussed above, serve as an illustration of such pooling as do self-help activities aimed at producing a common good. The costs of being ostracized tend to serve as the main enforcing mechanism.

Self-defense has always been a prominent feature of African societies and the norm has survived to this day largely because public law enforcement has been weak or unreliable. People have often been forced to take the law into their own hands. Vigilante groups have been organized to protect human life and, in rural areas, cattle and crops. Thus, external threats to the local community have served as powerful motivating factors to sustain such groups.

Charisma refers to a variety of heroes, saviors, prophets and shamans who are viewed as having extraordinary qualities and power. Such figures have always played an important role in Africa. They were critical to legitimizing traditional rule in precolonial society and have reappeared in modern times as part of the struggle against colonial rule and more recently as persons promising a measure of certainty and security in an increasingly globalized world. The many prophetic churches that have sprung up across the continent are one example of how charismatic leadership manifests itself.

These informal institutions constitute a vibrant part of African society. How far they permeate is hard to recognize because their presence is not always known. We know that the economy of affection helps to make sense of and legitimize market society. It generates the level of trust that shapes human interactions in the marketplace. Its face-to-face characteristics also shape the way bureaucracy works. Thus, the economy of affection and its informal institutions tend to have both positive and negative connotations.

The public tend to embrace it as something valuable while managers of modern organizations view it as undercutting the morality needed to pursue development at macrolevel. The formal and informal are at times pursued separately but where they meet each other, the outcome is never fully predictable because the formality associated with government institutions – what Ekeh (1975) referred to as the "civic public realm" does not enjoy the same legitimacy as the "primordial" public realm, i.e., the rules connected to the economy of affection.

To fully appreciate and understand the informal institutions of the economy of affection it is necessary to relax the assumption underlying neo-institutionalist theory (Williamson 1985; North 1990). Because social formations in Africa continue to rely on rudimentary forms of technology, social and political action is not collective in the sense that those familiar with modern industrialized society imply. People do not share goals that stem from having common economic interests. The division of labor in African economies has yet to produce rigid forms of social stratification that pitch workers against wealthy owners. The gap between rich and poor in Africa is there for everyone to see, but this does not mean that the predominant social cleavages reflect this inequality. The economy of affection and its informal institutions provide venues for the poor to receive benefits and often to move ahead on their own. Rich and poor do not perceive the world in terms of separate destinies. It makes sense, therefore, for people to seek improvement of their livelihoods by associating with others who share their culture and values. A shared identity based on membership in the same community of consumption rather than a shared economic interest based on divergent roles in economic production structures human exchanges in Africa and, by extension, society itself. The neo-institutionalist model of reconciling private preferences to achieve a public choice by ordering these preferences to achieve what amounts to a suboptimal outcome is typically not the way decisions are made in African countries. Issues are rarely resolved through compromise with a win-win outcome. More common is a form of resolution that involves the imposition of the

Table 3.1 Comparison of formal and informal institutions

Variable	*Formal institutions*	*Informal institutions*
Type of exchange	Impersonal	Face-to-face
Nature of exchange	Contractual	Non-contractual
Character of rules	Written	Unwritten
Approach to rules	Rule of law	Rules-in-use
Time schedule	Specified	Non-specified
Actor premise	Goal adherence	Shared expectations
Operative implication	Precise compliance	Ambiguous execution

Source: Author.

will of the most powerful actor. This is typically not a nakedly coercive act. Instead, it involves using various means of co-opting others to achieve consensus. Furthermore, there is always the expectation of reciprocity, the notion of giving back something, even if it is not the same exact value or is within a specified time. Breaking this code is costlier for decision makers than missing out on a consensual, suboptimal outcome. That is why most decisions are not experienced as "zero-sum" games in which the winner takes all. The differences between formal and informal institutions that follow from relaxing the assumptions of neo-institutionalist theory are summarized in the table below.

A comparative moral economy perspective

The economy of affection has grown out of Africa's Natural Society legacies to become a dominant feature of how livelihoods are secured and managed. The final question here is how it compares with other types of moral economy. Although it may be possible to trace morality in economics all the way back to Adam Smith, the concept in modern times tends to be associated with two names: E.P. Thompson and James C. Scott. Thompson (1971) recounts the history of the eighteenth-century food riots in the English countryside that erupted when some farmers, amid food shortages, decided to abandon the set price agreed upon by the local households in favor of selling at highest possible price in the market. The notion of a "fair" price was deeply rooted in the feudal system of the time and Thompson discusses how traditional norms of morality survive as a powerful social force against the intrusion of the market.

Fairness is also a centerpiece of Scott's analysis of the peasantry in Southeast Asia. He (Scott 1976) traces how the right to subsistence was invoked to guarantee the welfare of poor households but how state and global capitalism challenged these moral economy principles and exposed peasant farmers, especially those living close to the subsistence level, to greater risks that they must take into consideration in pursuing their livelihoods.

As we will discuss further in this volume, especially Chapters 7 and 8, there are notable similarities as well as differences between the situations in Southeast Asia and in Africa. The principle of reciprocity as the basis for forms of social insurance is there in both places but has been much further marginalized by colonialism and global economic forces in Southeast Asia than in Africa. The process of modernization in Southeast Asia has produced a degree of social transformation that is still to materialize in African societies. Much of this social transformation has been facilitated in Southeast Asia by the fact that countries in the region were organized into powerful states already long before colonialism. As we discussed in Chapter 2, except for Ethiopia, precolonial states in Africa were rudimentary creations and many societies were in fact stateless. Authority in the latter places did not rest with persons in an elaborate hierarchical order but with elders exercising power in specific lineage or clan structures. The colonial powers in Africa, therefore, encountered a much greater diversity of social formations that were not easily placed under a single roof for a coordinated push to modernize society. Furthermore, as we have discussed above, because these formations rested on rule over people rather than land, they were much more fluid.

Scott's analysis of the moral economy in Southeast Asia has much in common with Thompson's accounts of eighteenth-century England. Both emphasize how capitalism as a modernizing force poses a threat to the poorer segments of producers for whom falling below the subsistence level would have meant disaster. Their treatment of reciprocity and sharing, therefore, is viewed in the context of exploitation and marginalization while in Africa the economy of affection evolved in a historical context where the reach of the colonial state and capitalism was far less effective. The economy of affection, therefore, is more than just a social insurance mechanism. It has taken on a broader and more diverse role as parallel economy through which reciprocity may be applied also for obtaining goods and making a career outside of the local community. Boundaries of Africa's many communities of consumption today stretch over wide geographical territory. Compared with nineteenth-century Europe or Southeast Asia after the Green Revolution, these micro-institutions are not subordinated to market or state; instead, they shape much of the macroenvironment. This is quite unique in the contemporary global context. That is why understanding their roots in rural society is especially relevant and important. The next six chapters will provide a sense of the variety of ways in which non-agrarian features shape agriculture and why, as a result, the broader development scene in Africa needs to be the subject of rethinking.

4 Farming practices among African hunter-gatherers

Diversifying without loss of the past

Haruna Yatsuka and Kazunobu Ikeya

Introduction

Hunter-gatherers in Africa have traditionally been considered to depend almost entirely on natural fauna and flora, ranging from semi-desert to tropical rainforest. Formerly, this image of typical hunter-gatherers was also associated with the archetypal mode of life of pre-agricultural humankind. Such a view, however, has been criticized as a "traditionalist" standpoint, which assumes that there is an unchangeable essential quality in existing hunter-gatherer communities that can be utilized to reconstruct the past. Meanwhile, "revisionist" researchers have argued that the livelihood of present hunters and gatherers has been historically shaped by contacts with the outside world including cultivators and pastoralists as well as representatives of national political authorities. In the so-called "Kalahari debate" about what conclusions to draw about the human past from studying the present livelihood of the San people, revisionists claimed that such research would not tell much about how humans lived long ago (Ikeya 2002).

Earlier empirical studies to elucidate human-nature relationships and social organization, along with underlying ideologies among hunter-gatherer communities showed that they basically led a life "from hand to mouth" with no accumulation of wealth (e.g., Tanaka 1980; Ichikawa 1986). They repeatedly relocated, possessing only a limited number of goods due to their great mobility. They tended to allocate acquired foodstuffs equally among members of each band, devising various social measures to avoid concentration of wealth into the hands of a small number of individuals.

Today, many hunter-gatherers have become sedentary and accept farming and livestock husbandry. Categorizing their society as based only on subsistence activities does not make much sense since modern hunter-gatherers normally engage in a multifaceted livelihood strategy, flexibly combining (or shifting between) hunting, gathering, and various other activities, including occasional cultivation and livestock-keeping as well as wage labor, office work, trade, and even receipt of welfare benefits from the state (Bird-David 1992). This modernization of their livelihood system, however, does not necessarily imply a loss of their fundamental values. For example, Barnard (2002) shows how a "foraging mode of thought" persists even after their "mode of production" has changed, leading to the continued reproduction of Natural Society values such as the penchant for sharing of goods and immediate consumption.

This chapter highlights this set of issues by focusing on four groups with a deep-rooted tradition of hunting and gathering – the San of central Botswana, the Baka Pygmies of south Cameroon, and the Hadza and Sandawe peoples of north-central Tanzania. The first three communities still retain a vigorous hunting-and-gathering tradition, each living in considerably different natural environments, while the Sandawe have a longer history of farming. No group subsists any longer only on hunting and gathering but relies on farm products from either their own fields or specialist farmers. The intriguing question is what happens to these hunter-gatherer communities once farming is introduced into their quintessential Natural Society environment.

Farming practices and food sharing among the San

Contrary to their stereotypical image as "Bushmen," the San people in the Kalahari Desert have a long history of interaction with the outside world. As early as the nineteenth century, they had already been trading with neighboring peoples and foreign merchants and even offered local chiefs fur coats as a gift (Ikeya 2002). Archaeological and historical studies in the 1980s revealed that the San had been in contact with pastoralists and agriculturists nearly a thousand years ago (Wilmsen 1983, 1989).[1] At the same time, some of the neighboring Kgalagadi agropastoralists were assimilated into the San communities in the Kalahari, adopting hunting and gathering alongside their small-scale farming and livestock keeping (Ikeya 1999, 2002).

By the early 1960s the San had established a nomadic life based on hunting and gathering in the Xade area of the Central Kalahari Game Reserve (CKGR). This semi-arid area has an average annual rainfall of less than 400 mm, and surface water is available only 40–50 days per year in low grounds or pans. Tanaka (1980) revealed that more than 80 percent of their caloric intake came from wild plants gathered by women. The wild watermelon was particularly important as a source of water in the dry season. Tanaka also demonstrated that the San frequently relocated in response to seasonal changes in vegetation, and that their communities (small bands each made up of 30 to 50 people) were organized according to a strong egalitarian principle manifesting itself in the sharing of foods, especially game meat. The San depended on two species of wild watermelon; *//naan* (*Citrullus lanatus*) and *kaan* (*Acanthosicyos naudiniana*). While the *kaan* cannot be stored as they spoil quickly, the *//naan* can be stored and used for several months as a source of water in the dry season. As a result, they set up a camp where *//naan* are abundant. Once the fruit is exhausted, they move to other areas in search of plentiful *//naan* (Ikeya 2002: 75–80). In one nomadic camp, the number of stored *//naan* per household varied, ranging from 400 to 4,000. While some households short of *//naan* moved out to where *//naan* was available, others continued to stay in the camp, drawing on the generous distribution from *//naan*-rich households. In the San communities, the haves cannot refuse the demand from the have-nots (Ikeya 2007: 95).

The same principle of sharing was also observed in the case of cultivated watermelons. For example, a San who was able to harvest 1,000 watermelons in 1993 found people coming to demand their share of the fruits almost every day, with each of them being given four or five watermelons. They were mostly the farm owner's close relatives either by blood or marriage. Once they had received their share, the remaining watermelons were distributed to two neighboring non-relative households. In another case where a San harvested around 2,000 watermelons, six households set up their camps in the vicinity of the field on the expectation that they would receive the fruits since they had not engaged themselves in farming that season. Over a period of three months, the farm owner continued to distribute watermelons to his "guests," who were either his in-laws or friends. After the watermelons were finished in the field, four households broke camp and moved to other areas. This case shows that even non-farming households can claim a share of surplus watermelons cultivated by others. Therefore, the San who are skillful in farming find it hard to accumulate wealth from their harvest because, according to tradition, they have no option but to share their wealth with kin and friends (Ikeya 1996, 2007).

In 1979 a settlement meant to be permanent was built in Xade with the year-round provision of drinking water from a well (Ikeya 2018). A government office was established, and local officers began to promote agriculture and goat-raising, besides providing educational and medical services. People from different parts of the CKGR eventually settled in the vicinity of the Xade settlement, with its population reaching 530 by 1982. As consequence, wild edible plants around the settlement were quickly exhausted and the residents had to depend on either famine relief or impromptu farming (Tanaka 1986). By the mid-1980s, the settlement was still "semi-permanent" in the sense that households or individuals occasionally moved from one place to another, as they had in the past, although the frequency of movement was less (Ikeya 2002).

Ikeya closely observed the livelihood activities of the San, including their agriculture in Xade, from the late 1980s to the mid-1990s. The Xade residents at the time engaged in multiple livelihood activities, such as goat-raising, farming, hunting, gathering, craft making, and working at road construction sites. They flexibly combined these activities in accordance with the fluctuating precipitation pattern (Ikeya 1993, 1996, 2002).

The government attempted to promote agriculture through free distribution of seeds of maize, sorghum, and cowpeas as well as the introduction of ploughs pulled by donkeys (Tanaka 1986). Through the 1980s, the area of rain-fed farmlands around Xade increased. In the 1993 rainy season 37 households were engaged in farming. Farm size varied greatly, ranging from 0.1 ha to 2.5 ha, depending on whether the household utilized a steel plough. The size of the field could be flexibly expanded or reduced according to the amount of rainfall. Major crops were watermelons, cowpeas, and maize, which were normally mixed in various combinations in a single field. To protect farmlands from wild animals, fences were built around each field, laborious and time-consuming work. If the plants failed to sprout, the settlers would stop building fences and instead switch to hunting or gathering. Should there be an outbreak of edible green caterpillars, as happened in February 1993, some people would abandon farm work and start to gather this delicacy from the wild. Few settlers adopted protective measures against vermin (crickets and birds) that could cause damage to grain. All these facts suggest that farming had not established itself as a major source of food among the San villagers in Xade. For example, some people would abandon their fields after they received drought relief subsidies from the government (Ikeya 1996, 2002).

Eventually, the Botswanan government decided to remove all the residents from CKGR. Resettlement on a full scale started in 1997 with people being moved to permanent villages established by the government outside the game reserve (Ikeya 2001). In one of the new villages (New Xade or Kx'oenshakene), a member of staff of the Ministry of Agriculture was stationed to offer extension services and to provide seeds and ploughs. Kitchen gardens were promoted by an NGO, which distributed nets, seeds, and chemical fertilizers to grow a variety of vegetables. The San settlers continued farming, but their main crops remained watermelon and cowpeas, much like they had done before. They paid much less attention to other crops (Maruyama 2010: 92–3). According to the same author, who conducted research in Kx'oenshakene in the 2000s, among a variety of crops, watermelons are most likely to be shared among residents. Meanwhile, the area under maize cultivation is still small, and maize is rarely shared with others (Maruyama, personal communication, October 2018). In uncertain circumstances, falling back on sharing is a way of overcoming hardship, whether it is the result of climatic variation or unforeseen social reasons.

Farming practices and food sharing among the Baka Pygmies

Numerous studies have shown that symbiotic socioeconomic relationships have developed between the Pygmies and neighboring cultivators in the

African tropical rainforests (Bahuchet and Guillaume 1982; Takeuchi 2014). Generally, Pygmies supply bush meat or farm labor to the cultivators, who in turn offer farm products, tobacco, alcoholic drinks, or cash in return, as we discuss further in Chapter 5. These exchanges, along with pro-agricultural government policies, have over the years made the Pygmies gradually adopt farming to complement existing foraging activities in the forest.

By the 1940s the Baka Pygmies of southeastern Cameroon were leading a nomadic life, depending chiefly on wild plants and animals, besides obtaining farm products from neighboring cultivators in exchange for bush meat. Colonial and postindependence governments have since the 1950s encouraged the Baka to settle and farm. After motor roads were built in the 1970s and subsequent years the population began to settle along these new traffic arteries. Today, most Baka groups have set up permanent settlements in the vicinity of the villages of neighboring agriculturists (Kitanishi 2003: 145–7, 152). While the Baka are engaged in farming around these permanent settlements, they often continue to go deep into the forest, staying there for several weeks to two months to subsist on hunting and gathering. According to Yasuoka (2012: 100–2), who joined two-month long Baka hunting-and-gathering expeditions, major foods in the forest campsites include honey, bush meat, and wild yams. In the forest, there are some spots where annual wild yams (*Dioscorea praehensilis* and *D. semperflorens*) are concentrated. The Baka know very well how to find these spots where they can gather abundant wild yams as if they were in the "cultivated field" (Yasuoka 2011: 59). They intentionally leave a part ("head") of the tuber and rebury it to allow regeneration, an act which was called "para-cultivation" by Dounias (1993). It is likely that human intervention in forest vegetation encourages annual wild yams to thrive in certain patches, adapting themselves to human-made habitats in the forest. In this sense, these yam species are "semi-domesticated" plants, rather than genuine wild plants. Yasuoka (2012: 102) also observed that some Baka transplanted annual wild yams to their cultivated fields.

In the permanent settlements today, a major part of what the Baka eat comes from crops grown in their own fields, especially plantain, according to Kitanishi (2003) who made a survey of Baka agriculture in 1999 and 2000. At first glance, Baka farming practices appear to be similar to those of nearby cultivators. The Baka men clear either primary or secondary forest in the main dry season, burning dried woods and grasses at the end of the season. Women then plant major crops such as plantain, cassava, yautia, and maize. While the latter is planted all at once, the former crops are planted sporadically over a period of one or two years. Three or four years after clearing, they stop tending the crops and the fields are abandoned only to be cleared again after 15–20 years of fallow (ibid.: 147–8). The Baka farm is generally small in area and fluctuates from one year to another. The average farm size cleared annually per household was 0.27 ha in Kitanishi's case and 0.1 ha in Yasuoka's case, while neighboring cultivators annually cleared on average 0.74 ha (ibid.: 148–9; Yasuoka 2012: 91). It is obvious that agriculture is only secondary for the Baka who rather look at the forest as their prime source of food. Most Baka continue to hunt and gather, leading a "generalist" lifestyle according to which diversity rather than specialization counts.

The farming practices of the Baka are characterized by haphazardness and a lack of planning. They do not necessarily clear the fields every year, and even prepared fields are sometimes left unplanted and eventually abandoned. The major reason why they fail to clear the field or plant crops is that they are often away from the settlement in the farming season, either staying in the forest or visiting another village. Some Baka are busy working for neighboring cultivators, so they have no time to attend to their own fields. For example, during the short rainy and dry seasons in 2000, most Baka stayed in the forest for a long time to collect oil-bearing nuts (*Irvingia gabonensis*), one type of major wild foods that was especially abundant that year. As a result, many maize plants in their fields were left unattended and overgrown by weeds. Many Baka prefer the plantain because it does not need so much care and can survive even if the field is overrun with weeds (Kitanishi 2003: 148–50).

Kitanishi (2003: 150–1) argues that the haphazard agricultural practices among the Baka are associated with the "immediate return system" that characterizes hunting-and-gathering economies, in which a return from labor can be gained directly and immediately (Woodburn 1982). For example, game meat or edible plants are mostly consumed on the same day and shared with other members of the camp with no storage or accumulation. This is in contrast with the "delayed return system" which is associated with agricultural cultivation. The immediate return system tends to prevail where livelihood activities are heavily dependent on the natural condition of that very moment. Once the Baka get the information that some resources are available somewhere they may move to the forest promptly regardless of what they are doing, including for example preparing the crop field. As availability of natural resources such as animals and plants fluctuate from year to year, Baka themselves can seldom predict their own whereabouts even in the short run. An organized schedule for farming conflicts with the flexibility required in foraging activities. In addition, the Baka often consume all their harvested crops without keeping seeds for planting the following year, relying instead on nearby cultivators to provide seeds for planting. In this sense, farming practices have not brought about a real change in the Baka production system as their economy continues to be based on the "immediate return" system (Kitanishi 2003: 151). Kitanishi further asserts that by getting farm products from neighboring cultivators in return for daily labor, the Baka convert a delayed return system (farming) to an immediate return system (remuneration for a day's labor).

Another factor which tends to discourage the Baka from adopting full-scale agriculture is the constant social pressure to share the harvest with other members of the community. Among the Baka, only a small portion of crops is sold and almost all the farm products are eventually shared with others, mainly through a system of women sharing meals (Kitanishi 2000: 166). Sharing may also manifest itself in the "theft" of someone's crops in the field. This kind of behavior seems to be implicitly sanctioned by the community (Kitanishi 2003: 153). Yasuoka (2012: 106–7) also observed that the Baka not only harvest plantains from their own fields, but also sometimes "gather" from somebody else's farmland, as if they gathered plants from the wild. Even if it is not a fully acceptable behavior, it is rarely condemned explicitly. In their way of thinking ownership of foods is not necessarily associated with who invested labor to get the harvest. For example, according to Baka rule, the "owner" of a beehive is the first person to discover it. Interestingly, it is not the "owner" but other persons who do the laborious work of collecting honey. The laborers, however, do not have any priority rights in the process of distribution of honey, which is allocated equally to those who happen to be present. Thus, neither "owner" nor those who invested labor have full control of the disposal of acquired foods, a belief that stands in contrast to the thinking of the cultivator who firmly believes that ownership of a particular crop is vested in those who contributed their labor (ibid.: 107–9). The fact that some Baka villagers do not hesitate to "gather" plantains from the fields of others also indicates the vague demarcation that exists between the cultivated field and the wild. Yasuoka (ibid.: 109–10) suggests that for the Baka, plantain cultivation is a semi-domesticated activity much like the gathering of annual wild yams, which is viewed as something in-between collecting genuine wild plants and full-scale farming.

The Baka have come a long way from being pure forest dwellers. They have adapted to a life in which their sources of livelihood have become increasingly diverse. Yet, they have retained much of their Natural Society outlook. The question that has arisen more recently is how the Baka are responding to the recent introduction of cacao. Being a perennial tree crop, cacao may encourage the Baka to consider it an inheritable property. Oishi (2012), for example, reports that 90 percent of all the Baka households in his study area had at least one cacao field. He also notes that it is common that the cacao fields are inherited over two to three successive generations, which shows that the land producing cash crops may be regarded by the Baka as inheritable. This may suggest that an element of the delayed return system is entering their livelihood strategy and they may be taking their first steps toward becoming landed farmers (ibid.). However, Oishi argues further that the Baka still grow cash crops in line with the immediate-return principle in the sense that they normally sell all harvested cacao without waiting for the price to rise in the post-harvest season and immediately spend whatever cash they earn.

Livelihood diversification and food sharing among the Hadza

Studies of the Hadza hunter-gatherers in northern Tanzania have been mainly conducted in the most remote parts of Hadza-land. Over the years these peoples have had the least contact with other groups of people practicing agriculture or keeping cattle (Blurton-Jones 2016: 119). Therefore, little information about their livelihood activities other than hunting and gathering has been available. In fact, attempts to transform the Hadza from mobile foragers into settled farmers have been made several times by both the colonial and postindependence governments, but with little success. Instead they have chosen their own path taking advantage of local opportunities as they arise.

The Hadza people today are engaged in hunting, gathering, tourism, and agriculture. How these activities are combined varies from one settlement to another and changes flexibly as each individual household relocates (Yatsuka 2017). Today, farming is often practiced in these communities, particularly among the residents of settlements established in the 1990s. They cultivate sorghum and maize for self-consumption relying mainly on rainfall although it is too low (approximately 500 mm per annum on average) for practicing stable rain-fed farming. In stark contrast, in Mang'ola, the eastern part of the Hadza's inhabited area, neighboring cultivators grow maize and onions on a commercial basis using irrigation with spring water. In the same way as the Pygmies and the San, the Hadza do not necessarily engage in farming every year, instead freely choosing other livelihood activities.

Ethnic tourism to experience the Hadza's hunter-gatherer lifestyle started in the 1990s and has since become popular. As a result, some of their camps in Mang'ola are visited by foreign tourists almost every day. The Hadza earn income from entry fees paid to each camp at a fixed rate as well as individually from selling souvenirs (Yatsuka 2017). It is not uncommon for the Hadza to move to areas active in tourism during the dry season to earn their living from doing business with tourists. As the rainy season starts, some of them would like to move to another settlement to farm but not every Hadza engages in farming and many prefer to obtain maize, their staple food, from neighboring peoples, either with the cash they earn from tourism or in exchange for bush products notably honey. In the Mang'ola camps, the residents purchase maize with the cash gained from tourism business, which is then shared by all the members of the camp. In another area, a Hadza man explained that the sharing system of harvested crops differs from that of food obtained from hunting and gathering. According to him, bush meat is shared and eaten together, but they do not share harvested grains, which are stored in each individual house. This remark suggests that, unlike perishable crops such as watermelon (San) or plantain

(Baka), harvested grains are unlikely to be shared with others, possibly because they can be stored.

The history of farming practices among the Sandawe

The Sandawe of north-central Tanzania have a relatively longer history of farming and animal husbandry than the San, Pygmies, and Hadza. Most Sandawe adopted agriculture and had already begun to raise livestock in the late nineteenth century through their interaction with neighboring ethnic groups (Newman 1970; Yatsuka 2016). Unlike the previous three case studies, farming practices seem to have spread spontaneously among Sandawe without any significant government intervention. Today, agriculture is the mainstay of the Sandawe subsistence economy and not just one of many options to secure foods. Hunting and gathering, however, are still important. Both activities play an essential role in their everyday life as our comparison below with data collected by Newman (1968, 1970) illustrates.[2]

Residential pattern and livelihood until the mid-1960s

In the 1960s, the Sandawe lived in small settlements in the bush, made up of a few households related to each other through paternal kinship. Since they cultivated fields around their houses and in their neighborhoods, they often relocated not only their farmlands but also their houses to a new place once they found that the existing farmland was exhausted and there was no additional land to cultivate. Newman (1968) noted that government officers were always frustrated by the difficulty in organizing the Sandawe people into a community, as they stayed at a particular place for only a short period of time.[3] Their land tenure system was characterized by a lack of sense of ownership, with people shifting their farmlands freely and frequently. There were no clear markings for land boundaries among the Sandawe. If a piece of land was available, even people from other ethnic groups could be allowed to use it without seeking permission from local clan elders (Newman 1970: 39).

Although domestic animals are considered to have been introduced into Sandawe society earlier than crops, they owned only few domestic animals in the 1960s. Meanwhile, the Sandawe grew a variety of crops, including pearl millet, sorghum, maize, pumpkin, sweet potato, common bean, Bambara groundnut, and cowpea (Newman 1970). Groundnuts and castor-oil plants were cultivated as cash crops. Research on the Sandawe diet conducted by Newman (1970) revealed that maize was becoming the main staple for them at that time, partly replacing pearl millet and sorghum.[4] The immediate cause of this change might be the severe famine in 1961–2. Maize as famine relief was imported from the United States and distributed across north-central Tanzania. Maize seeds were the only ones available for sowing the following year. Maize also gained popularity among the Sandawe because it was less susceptible to bird attack than pearl millet and sorghum. In addition, the spread of grain mills encouraged them to adopt the cultivation of maize, which was hard to grind using only a locally made mortar.

Hunting and gathering were still important livelihood activities in the mid-1960s. Men went hunting several times a year, using such tools as nets, bows and arrows, and traps. According to Newman's survey, some kinds of animal products were eaten on 54 percent of the surveyed dishes accompanying their staple diet (a firm porridge known as *ugali*), with 9.4 percent being bush meat or fish. They also often consumed wild fruits and leafy plants. According to the same survey, 45 percent of the dishes were made from leaves or seeds of wild plants or wild mushrooms (Newman 1980). Sandawe men also frequently collected honey either from man-made or wild beehives.

Residential pattern and livelihood in the 2000s

Not long after Newman had completed his fieldwork, a compulsory *ujamaa* villagization program was implemented by the postindependence socialist government. Hitherto dispersed populations were forced to move into pre-designated roadside villages, with modern facilities, including churches, elementary schools, and government offices. High concentration of houses in a compact area made it difficult for the Sandawe to have access to enough farmland near their homes. As a result, some sought land for cultivation far away from their residence, while others made do with nearby fields that were less fertile. According to a survey made by Yatsuka in 2007, more than 30 years after *ujamaa* villagization, 30 out of 41 fields were located less than two kilometers (a 30-minute walk) away from their residence. The size of each field was in direct proportion to the distance covered from the house (Yatsuka 2012). This clearly shows that to be able to cultivate as large an area as they did before villagization, villagers had to walk back and forth a long distance from their current residence.

Today shifting cultivation with frequent relocation is no longer feasible but villagers still rotate their fields in an approximately five-year cycle, although some cannot afford such a long fallow period and instead use their land on a semi-permanent basis. They mainly grow grains, including maize, pearl millet, and sorghum, along with legumes, cucurbits (gourds), pumpkins, cassava, and sweet potatoes. According to Yatsuka's research in 2004, the fields producing staple grains accounted for approximately 90 percent of the total area under cultivation. All these cereals are used not only to cook their staple food, *ugali*, but also to make local beer, which is indispensable to their cooperative farm work (ibid.: 52).

Aside from the widespread practice of mixed-cropping, grain production itself is characterized by polyculture of a "generalist" nature. Table 4.1

Table 4.1 Combination of grains among 34 households (2007)

Pearl millet	*Sorghum A*	*Sorghum B (for alcohol)*	*Maize*	*Sorghum C*	*Sorghum D (improved)*	*No. of household*
+	+	+				5
+	+	+	+			4
+	+					3
+		+	+	+		3
+	+	+	+	+		2
+		+				2
+			+	+		2
	+		+	+		2
+	+	+	+		+	1
+	+	+		+		1
+	+		+	+	+	1
+	+		+			1
+		+	+	+	+	1
+			+		+	1
+						1
	+	+	+	+		1
	+					1
		+	+	+	+	1
		+				1
Total						34

Source: Yatsuka (2012: 54).

shows the combination of major grains for 34 households in one village (labelled Village F) in 2007. These households grew six crop varieties in total, and as many as 19 different combinations. Out of the six varieties, most households preferred pearl millet, maize, and sorghum A and C for making their staple food, *ugali*. Pearl millet is the most favored grain, as it has traditionally been used for multiple (both culinary and cultural) purposes. Pearl millet, sorghum B, and maize are also used for making local beer. This tendency of cultivating multiple grain varieties rather than specializing in one type of grain can be widely observed across the Sandawe territory. Such diversity not only contributes to enriching their food culture but may also help them to cope with an unstable semi-arid climate.

For their everyday dishes other than the staple *ugali*, the Sandawe are greatly dependent on wild products including leafy plants, mushrooms, and bush meat, among other things. It is notable that some wild plant species growing in their crop fields constitute an integral part of their diet (Table 4.2). Among these "edible weeds", false sesame (*Ceratotheca sesamoides*) is the most important. These weeds are easier to gather in the fields than in the wild. Once they have been collected in the rainy season, they are either eaten fresh on the same day or sun-dried for the dry season when fresh leaves for consumption are difficult to find. According to a

Table 4.2 "Edible weeds" in Village F

Family	*Binominal name*	*Edible part*
Pedaliaceae	*Ceratotheca sesamoides* Endl. ("false sesame")	leaves
Pedaliaceae	*Sesamunm angustifolium* (Oliv)	leaves
Tiliaceae	*Corchorus trilocularis* L.	leaves
Capparidaceae	*Cleome hirta* (Klotzscg.) Oliv.	leaves
Amaranthaceae	*Amaranthus graecizans* L.	leaves
Amaranthaceae	*Amaranthus hybridus* L. ssp. *cruentus* (L.) Thell	leaves
Amaranthaceae	*Amaranthus hybridus* L. ssp. *hybridus*	leaves
Convolvulaceae	*Ipomoea farinosus* L.	leaves
Aizoaceae	*Gisekia pharmaceoides* L.	leaves

Source: Yatsuka (2012: 171).

dietary survey during the 2006 rainy season, wild plant leaves and mushrooms accounted for 52 percent of the total number of dishes. Of this, 38 percent were "weeds" from the crop fields. In another survey conducted in 2004, more than half of the dishes in the dry season were made from dried leaves of false sesame (Figure 4.1). Thus, Sandawe farmlands not only produce crops but also considerable amounts of "edible weeds", which are neither included in agricultural statistics nor regarded as "farm products" by agronomists. As the villagers intentionally encourage the growth and reproduction of these species, they can be regarded as living in the "semi-cultivated" status.[5]

Today, the Sandawe have developed diverse livelihood strategies, including expanded commercial crop and livestock production. Although hunting and gathering are not considered as important as they were in the 1960s, some villagers are still active in foraging for both subsistence and commercial purposes. Wage labor is an important option to make a living particularly in years of crop failure. Compared to their past, the Sandawe today are increasingly becoming "generalists," spreading their options to earn a living in their uncertain semi-arid environment. The relative

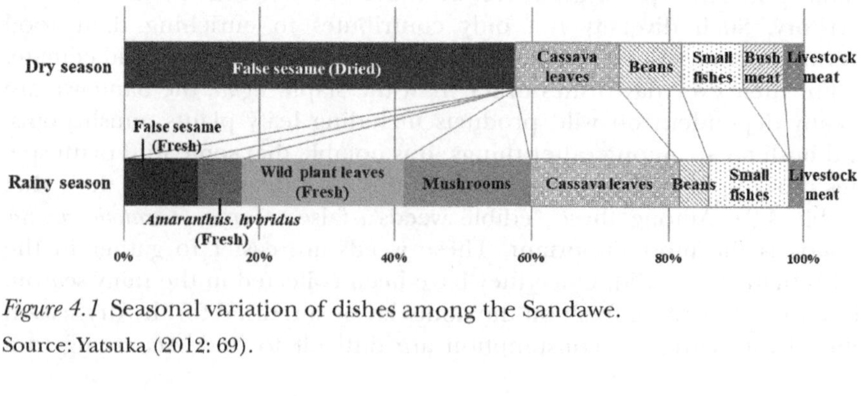

Figure 4.1 Seasonal variation of dishes among the Sandawe.

Source: Yatsuka (2012: 69).

importance of each livelihood activity (and of each crop) differs from one household to another, but these differences tend to be evened out through sharing and exchange of foods between villagers.

Food sharing among the Sandawe

Like other rural dwellers in Africa, the Sandawe often share a cooked meal with others who happen to be there. Unlike among the San and the Baka, however, the ethos of sharing is not so strong among the Sandawe villagers. Although they may give others wild products they have gathered (for example, the leaves of false sesame), they rarely share the crops (especially grain) harvested from their fields, the exception being the calabash, which is often shared with neighbors and friends as they ripen in large quantities at once.

False sesame, the most important edible weed, is sometimes collected from the fields owned by other villagers (mostly relatives), by asking permission from the farm owners, but like the case of plantain among the Baka, it might be "gathered" from the field of others without permission. Of course, gathering something from other people's fields without notification is not encouraged and is not openly spoken of, but it does not lead to overt conflict unless it happens too often. Even if there is a complaint, the gatherer would not be accused of "stealing" because false sesame "just grows naturally" (Yatsuka 2012: 174–7).

The diversity of crops grown at household level, as well as the availability of other kinds of food, is eventually balanced by a barter exchange system called *tele kwa tele* in Kiswahili. Meaning literally "full by full," *tele kwa tele* refers to exchanging one full container of something for another full container of something else, using the same size of container to measure the volume of each product, regardless of its weight and market value. By using this barter exchange system, a villager can trade his own grains for other varieties. Many other food types are also exchanged using *tele kwa tele*, including dried edible weeds, mushrooms, insects, and beans. For instance, one liter of pearl millet can be exchanged for one liter of dried leaves of false sesame. *Tele kwa tele* thus enriches the diet of villagers by making additional food items available for each household.

Conclusions

By examining the farming and food sharing practices among four different peoples with a long-established hunting-and-gathering tradition – the San, the Baka, the Hadza, and the Sandawe – we have seen that despite government efforts to turn them into "specialist" farmers or herders they have stuck to their "generalist" subsistence regime. They continue to combine a variety of food-producing and income-generating activities, not only hunting and gathering, but also farming, goat-rearing, wage labor, tourism, and so forth.[6] Farming remains only one option, except for the Sandawe, which is the only group that has a long tradition of practicing agriculture. Even they, however, seek diversity as their unstable agriculture in a semi-arid environment is full of risks and thus tends to be supplemented by a flexible system combining several different subsistence pursuits. This diversity of livelihood strategies brings about disparity among households in terms of access to certain food resources, but it is mitigated by the practice of sharing in the cases of the San and the Baka or barter exchange in the case of the Sandawe.

There is no convincing evidence that these groups of people are in a transition from being hunter-gatherers to becoming agriculturalists, as Yasuoka (2012: 86, 109) argues in reference to the case of the Baka. Farming practices among the San and the Baka are considerably different from those of cultivators, as agriculture has yet to take firm hold in their livelihood system. Farming was accepted as a part of their livelihood strategy of resource use diversification. Another important point is that there is continuity between Baka farming practices and their utilization of forest products, as can be seen in the cases of semi-domestication of wild yam species and haphazard plantain cultivation. It is difficult to make a clear demarcation between cultivation and gathering, and between crops and "semi-cultivated" plants. This is true also for the Sandawe despite having been engaged in farming for a long time.

A special challenge to many hunter-gatherers is that they are living in areas with rich natural resources, which means that their habitats have in many instances been designated as wildlife management areas. For example, all four communities studied in this chapter are now restricted one way or the other as they engage in hunting. Furthermore, as they have become sedentary, concentration of population naturally leads to an acute shortage –as well as rapid degradation – of land. This issue has been behind government schemes to increase efficiency of land use by turning hunter-gatherers into full-time farmers or herders. This bias is evident, for example, in the fact that wildlife conservation officers in the San resettlement areas are dubbed "do just one thing (*/kui sa kuru*)", as they always try to persuade the San to do just one thing (e.g., keeping cattle) instead of hunting and raising stocks at the same time (Maruyama 2010: 140). Incomplete and inefficient farming practices (such as among the Baka) may in the long run have to be replaced by a more specialized way of agriculture. Such specialization, however, is very likely to bring about food insecurity among the communities of hunters and gatherers.

There is a need, therefore, to find a more sensible and workable approach, in which their inherent socioeconomic flexibility and resilience can be ensured, and the copious indigenous knowledge of local communities can be utilized. In promoting agriculture in these communities, it is essential to consider the place of farming in their total system of subsistence, which differs considerably from one place to another or even from one household to another, according to the geographical as well as socioeconomic environment in which they find themselves. For example, for the San, who are now settled in densely populated villages outside a major game reserve, there is a need to establish a viable livelihood pattern suited to the dryland conditions which can support human lives without recourse to government food rations. Such a policy would have to take into consideration the persistence of hunter-gatherer values, notably the ways they use their natural environment and retain a light footprint by prioritizing the value of diversification.

Notes

1 Studies in genes have recently revealed that hunter-gatherers have had a long history of intermarriage with cultivators in neighboring communities (for example, see Tishkoff et al. 2007; Patin et al. 2009).
2 The area in which the Sandawe reside encompasses several villages in Chemba District, Dodoma Region, central Tanzania. The area is mostly hilly with an altitude of 900 to 1,500 meters. The first author conducted fieldwork in Village F, located in the southeastern part of the Sandawe territory. The annual rainfall in the study areas is approximately 600 mm, with a significant yearly fluctuation.
3 When the first author asked elderly people about the reasons for frequent relocations in the past, most answered that they moved in search of new farmlands for swidden cultivation, or, because "the house grew old." On the other hand, none answered that they moved in search of better places for hunting and gathering.
4 According to the 1950 census for central Tanganyika and the Kondoa District, the production of maize at the time was very small. However, according to data from the 1965 census, maize ranked highest in terms of both cultivation area and production (Newman 1970).
5 Consumption of "edible weeds" (including false sesame) is also widespread among the neighboring ethnic groups including the Gogo agropastoralists.
6 In the case of the San, government provision makes up the great bulk of the supply of maize flour in the settlement areas (Ikeya 2002: 22).

5 Shifting cultivators as generalist managers

The threat of "saving the rainforest"

Kenta Sakanashi, Kazuhiko Sugimura and Tadasu Tsuruta

Introduction

Mainstream discourse on "primitive society" makes a clear distinction between societies of hunter-gatherers and peasants, the latter being placed at a higher level on the evolutionary ladder of humankind. For example, Meillassoux (1967) contrasts hunter-gatherers with peasant cultivators in Africa by highlighting conflicting social attributes associated with each livelihood pattern. While hunter-gatherers lead a nomadic life, peasant cultivators are sedentary, bound to the land they till for survival. While the former only temporarily and intermittently cooperate with each other, the latter have a regular system of cooperative work. In social organization and land use, according to this perspective there is an impassable gulf between communities of hunter-gatherers and peasants, the latter eventually evolving into stratified market societies.

A prominent feature of the Kyoto School has been to call this perspective into question by demonstrating the interlinkage between different types of livelihood systems in Africa. By analyzing rural Africa in the context of the grand historical evolution from apes to *homo sapiens*, the early ecological anthropologists in collaboration with their fellow primatologists, as discussed in Chapter 2, focused on the elements of Natural Society among rural dwellers in Africa. They did not make a clear demarcation between hunter-gatherers, pastoralists, and cultivators. To them, what these societies have in common as a result of interacting with each other and their natural environment became as important as how they differed. Hence, their emphasis lay on such factors as geographical mobility, communal ownership of property, and propensity to egalitarianism.

By focusing on shifting cultivators in rainforests and savanna woodlands, this chapter will discuss how peasant communities in Africa still retain the elements of Natural Society. It begins with identifying general features characterizing shifting cultivators in Africa, by using chiefly ethnographies written by the first generation of the Kyoto anthropologists. It then examines the changing process of contemporary shifting cultivator societies, based on an empirical case study of a cacao-producing area in southern Cameroon, where Sakanashi carried out detailed field research. It shows how the introduction of cacao as a major cash crop into the peasant communities affects their relationship with neighboring hunter-gatherers. It ends with a discussion of how recent government policy to conserve the natural environment threatens grassroots livelihoods in the rainforest.

Subsistence agriculture and peasant communities in Natural Society

Subsistence is the core principle driving agricultural production in Africa's Natural Societies. It takes place in an environment that is little changed over time and, above all, is adjusted to what circumstances permit. Income from land, however, has become increasingly insufficient as local peasant communities have become integrated into the market economy. The response to these changes has been a diversification of livelihood sources rather than an intensification of land use in search of higher productivity. Yet another feature of these Natural Society communities is the tendency to share of whatever household members may have among themselves, so nobody goes hungry. We shall cover each of these points in greater detail below.

Shifting cultivation embedded in a natural environment

Shifting cultivation, variously referred to as "swidden" or "slash-and-burn" agriculture, also "forest (bush) fallow", has been a dominant form of farming in sub-Saharan Africa for a long time. This is still practiced widely, particularly in humid forest zones with an annual rainfall of more than 1,500 mm. Mixed cropping, also known as intercropping, in which more than one crop is grown simultaneously on the same piece of land is also a widespread practice across the subcontinent. The combination of forest fallow and mixed cropping (or just "fallow-mixed cropping") is a farming system embedded in the forestry ecosystem, in the sense of being dependent on the natural cycle of vegetation regrowth and the replenishment of soil fertility. This system stands in contrast to the major farming systems of Agrarian Societies, which are characterized by humanity's determination to control nature.

The fallow-mixed cropping system has generally been considered a rudimentary technology characterized by low levels of land productivity. In addition to being viewed as incompatible with commercial production, it has been criticized for leading to extensive deforestation. In a counter-argument, Paul Richards (1985) argued that both shifting cultivation and mixed cropping are rational farming systems with their own advantages, bolstered by villagers' intimate knowledge of the ecosystem and their constant innovation. African indigenous farming systems, therefore, have an untapped potential for development. Mixed cropping has its own merits by deconcentrating risks of disease and pest, securing foods throughout the season, providing a variety of dietary nutrients, and levelling uneven labor peaks, as harvest seasons for each crop tend to differ.

Such a farming system is based on the cultivator's detailed knowledge of the ecosystem. Different names are often applied to fields at different stages of cultivation and fallow. Suehara (1983: 15–16) found that the Tembo people living in the mountainous forest of Zaire (now the Democratic Republic of Congo) classified swidden fields into the following five sequential stages: (1) *mbaka,* the first field in rotation that is cleared from the virgin or secondary forest in order to plant maize, cassava, and gourds; (2) *chibuba,* the second stage field when cassava stays as the sole crop for two years; (3) *mtendo,* the third stage field left in fallow for three to five years; (4) *miyaka,* the first-year field cleared at the *mtendo* stage when all kinds of dominant crops are planted; and (5) *chitokolo,* the field abandoned after use as *miyaka* two to three times.

The fallow-mixed cropping system is closely associated with the communal system of land tenure. In most communities of shifting cultivators, land is basically controlled by the lineage group, and individual villagers are allotted pieces of land to cultivate, having only a usufructuary right. In the case of the Tembo, the land is controlled by each lineage group, and uncultivated lands (*mtendo* or *chitokolo*) must be returned as a common property of each group.

Generalist managers of multifaceted subsistence activities

Based on his empirical study among the Tongwe people in western Tanzania, Makoto Kakeya showed that their livelihood was marked by a "generalist" attitude to utilizing the natural environment. They engaged not only in farming but also in other livelihood activities such as hunting, gathering, and fishing (Kakeya 1986). Unlike "specialist" farmers in Agrarian Societies, who may be engaged in hunting and gathering only as "minor subsistence" activities,[1] African peasant cultivators in many places depend on the wild for a significant part of their diet. In Tongwe society, for example, everyday dishes were prepared with fish, bush meat, insects and/or wild edible plants (Kakeya 1976). In the forest areas of central Africa, hunted animals are particularly important as a source of protein.

The generalist character of shifting cultivation can also be observed in how villagers allocate time to various activities. Kimura (1992), who carried out his research among the Bongando, also in the former Zaire, shows that male villagers spent more time on hunting than on agriculture, referring to them as "multi-subsistence people" rather than pure cultivators. Based on his study of the Kumu in the same country in the 1980s, Sugimura observed that there was no trade-off between farming and hunting activities, and that livelihood patterns had been diversified with the emergence of other income-generating activities, including mining gold and diamonds (Sugimura 2004: 280–1). Sugiyama (1988) recorded time allocation in detail among the Bemba in northern Zambia for a total of 267 days in a ten-month period, during which she found that on average, a man spent 58 days (22 percent) on farming, 48 days (18 percent) on hunting and gathering, and 66 days (25 percent) on activities outside their home village (chiefly peddling goods).

The economy of sharing

As we indicated in Chapter 2, Kakeya (1976) highlighted how the Tongwe shifting cultivators tend to engage in what he calls a "minimum subsistence effort," i.e., an unwillingness to expand production beyond their minimal needs whether in farming or hunting and gathering activities. Villagers cultivated only enough acreage to assure their subsistence and gathered (or hunted) wild produce only in immediate environs, not embarking on hunting and fishing expeditions to remote areas. Another tendency he observed was the "equalization of consumption," a custom according to which inequality in household food stock was levelled through frequent visits with each other at which the guests, who may be lacking food, must be treated hospitably with a proper diet. Because of the obligation to share, villagers lacked the incentive to produce a surplus. The wealthy lived in constant fear of envy from others, who might put a curse on them if they refused to share their wealth. Thus, the Tongwe mode of production based on a minimum subsistence effort cannot be separated from their mode of consumption and its "levelling mechanism" underpinned by supernatural beliefs (Kakeya 1986).

Sharing labor is another important feature of agricultural production in rural Africa. For example, among Tembo villagers in the 1980s family labor dominated but cooperative exchanges of labor were practiced to complete demanding activities such as clearing bush, sowing seeds, and harvesting. This cooperative labor among the Tembo is divided into three types. *Likilimba* is a labor exchange system in which the same kind of effort is reciprocated between individuals. For example, a large-scale *likilimba* may gather around 20 females, who are treated to a meal and banana beer after the work is completed. The event is marked by a festive atmosphere with singing, dancing, and feasting. Sometimes even women with no land to cultivate take part simply because they want to participate in the cheerful gathering. *Lukoo* is a unilateral donation of labor without return, in which young males in their twenties and thirties help poorer old men who, in return, provide only plain dishes. Without the *lukoo* system, these old men would not be able to carry out farming, especially the heavy task of clearing bush. *Kwanza,* finally, is carried out by relatively rich (often polygamous) men. It involves preparing meals with goat meat and banana beer in exchange for labor provided by other men. Although the market prices of goat meat and banana beer are much higher than the wages for hired labor in this area, paying wages only would not be enough. With providing a costly feast of goat meat come prestige and power. For the Tembo, the wealthy are not men who control land or livestock but those who control labor (Suehara 1983).

Livelihood patterns of the Fang in southern Cameroon[2]

The Fang is an ethnic group living in the rainforest located in southern Cameroon (Figure 1.1, p. 14), where population density is still considerably lower than that of the other parts of the country.[3] The study villages are located along the trunk road linking local towns. The Fang people live side by side with the Baka, formerly known as just Pygmies, who had originally been nomadic hunter-gatherers, but who are now semi-sedentary and also engaged in small-scale farming, as discussed in Chapter 4. Because the Fang also frequently hunt wild animals, it is difficult to affix a label such as "(pure) cultivators" or "(pure) hunter-gatherers" to either ethnic group. The Fang and the Baka have a close relationship with each other through farming and hunting activities.

The Fang produce their staple foods such as cassava, plantain, and maize in swidden fields, which are categorized according to three different stages of land use. Uncropped virgin land or land after a long period of fallow (as extensive as several decades) is called *efane.* After clearing *efane,* plantains are planted, and then fallen trees are burned with special attention so as not to damage the planted crops. After harvesting plantains for three to four years, the field is left fallow for several years (*ekoro*). When the field is eventually cleared, trees are burned before cassava, peanuts, and maize are planted. Farmland under cultivation is called *afup.* Various crops are mixed in the same field, and every household always keeps two or three different fields (around one hectare each) in different stages of crop growth, which enables them to obtain foods continuously throughout the year. Use of farmland is based on lineage and each household has only a usufruct right so long as the field is under cultivation. Although plantain and maize are occasionally sold, the major cash earner in the study area is cacao. In the study villages, 61 percent of households possess cacao fields, and, on average, they earn 70 to 80 percent of their annual income from cacao.

Hunting and gathering are also important activities for the Fang, carried out throughout the year regardless of the season. Both trapping and gun-hunting are practiced, with the main game species being the blue diker (*Caphalophus monticola*) and the African brush-tailed porcupine (*Atherurus africanus*). For the Fang, hunting has multiple purposes. Traps are set around farmlands in order to prevent damage to crops by wild animals. Traps are also placed deep in the forest to catch animals for consumption or sale. In the case of gun-hunting, a Fang villager may be accompanied by a Baka, who has a copious knowledge of wild animals and often undertakes the task of shooting them. Game meat, the favored dish and the most important source of protein, is not only consumed domestically but also sold especially to urban folks whose demand for bush meat has been growing. Bush meat is also important to mobilize necessary labor in farming, as discussed further below.

The making of palm wine, which is collected from fallen oil palms, is another important activity involving mostly males. Oil palms, as a cultivated species gone wild, can be found extensively in the fields and secondary forests. Villagers often cut down an oil palm, gathering sap oozing from the tree and then fermenting it. From one tree, it is possible to gather two to three liters of sap a day for three or four consecutive weeks. Palm wine is mainly served at social gatherings such as daily spontaneous drinking occasions, parties after farm work, or treats for visiting guests. If villagers get a lot of palm wine, it is distributed to friends, family members, and even outsiders. The sharing of palm wine contributes to cementing and establishing social relations between Fang villagers as well as between members of different ethnic groups, as it offers an occasion to gather and talk over the beverage. Palm wine is also an important source of energy, vitamins, and minerals. Edible mushrooms, nuts and berries, along with herbs and bark are collected and used as condiments or medicines. Bush meat, palm wine, and other wild produce are not only important as income source for those with a small cacao field but are indispensable also for villagers with large cacao fields who need to organize festivities to attract extra labor.

Cacao production among the Fang and their neighbors

Traditionally, the Fang are considered to have led a nomadic life, depending on swidden cultivation, and hunting and gathering. Their society was acephalous, and they frequently relocated themselves in small bands made up of several households. Cacao was introduced into this area in colonial times and the Fang abandoned their hitherto nomadic lifestyle and became sedentary, in part due to the nature of cacao as a perennial tree crop (Bennett 1898; Alexandre and Binet 1958; Joseph 1977). Cacao production in southern Cameroon has generally been on a small scale, mainly carried out by indigenous forest dwellers, in contrast to other cacao-producing areas like Ghana and the Ivory Coast, where migrants were instrumental in expanding commercial cacao production (Hill 1963; Roussel 1971).

Although cash crops may also eventually become food crops – especially true for maize – cacao is a pure cash crop in the sense that it is not consumed by the locals. Only its pulp is occasionally consumed as a snack. Studies of cacao production and shifting cultivation in southern Cameroon indicate that villagers don't expand their land acreage and instead choose which crop to plant on what piece of existing land. Thus, when villagers specialize in cacao production, swidden fields (for food production) decrease, and when they concentrate on swidden fields, the cacao fields tend to be abandoned (Ndoye and Kaimowitz 2000; Sunderlin et al. 2000). This is in contrast with what Hill (1963) found in Ghana, where those who actively invested in land primarily expanded their cacao farm.

Even though the acreage of cacao production in southern Cameroon has not really increased very much, it is now well integrated into the existing system of production. The existing division of labor between sexes may have helped to facilitate the acceptance of cacao production. Traditionally, women managed the swidden fields for food production and, except for the hard work of clearing bushes, men were largely engaged in activities other than farming including hunting. When cacao was introduced, however, men with much spare time took charge of its production. In addition, the neighboring Baka, most of whom have no cacao fields, proved to be helpful in relieving severe labor shortages in cacao production.

The ecology of production

Most cacao trees in the study area were planted two generations ago, the earliest being more than 50 years old. Because cacao is a perennial crop it occupies a particular piece of land for several decades, at least as long as the field is well looked after.[4] A number of tall trees usually remain in the cacao fields, particularly useful species for medicine or condiments, providing shade cover for the shorter cacao trees. Oil palms, avocado, and citron may also be found in the cacao fields. Cacao seedlings are often planted alongside a plantain, in order to protect the cacao from sunlight. In accordance with vegetation regrowth, this plantain field is gradually turned into a cacao field indicating how cash crop production and food production are integrated into the same farming system.

A similar situation has been observed in southeastern Cameroon. In her detailed agroecological study, Shikata (2013) reveals that cacao production among the Bangando, the local ethnic group, is well adapted to the existing system of forest-fallow cultivation in that cacao and other food crops coexist in the same field. The Bangando farming system is based on the cycle of several years of harvesting food crops (especially plantains, their staple food) and more than ten years of fallow. The principle of their food production is to harvest plantain continuously throughout the year by keeping several different plots (and individual plants) at different stages of maturity along with mixing different cultivars. New swidden fields are repeatedly cleared and the old fields are abandoned to become secondary forests which continue to produce plantains for some years. Thus, a mosaic of different vegetation is created, each of which provides useful wild animal and plant species specific to each stage of ecological succession. Interestingly, plantain is also harvested from cacao fields, which are normally created by clearing pristine forests near rivers. Early cacao fields are also the places for plantain production but as years go by they are transformed into purer cacao fields while still providing space for various other useful tree species. Although cacao in good condition can be harvested for anywhere between 30 and 50 years, cacao fields may be abandoned, or overgrown by other plant species, if for some reason the necessary weeding is not done. There are a variety of reasons for abandoning cacao production such as shortage of labor, crop disease, and declines in the cacao price. Therefore, cacao production among the Bangando is closely integrated into their swidden cultivation system in which food production is assigned the highest priority and cash crop production can always be abandoned in favor of other subsistence or income generating activities.

Sharing of labor and food

Both food and cacao production among the Fang often require extra labor from outside of the household. Nowadays, many young villagers are away in urban or peri-urban areas, either for their education or wage labor. In Fang villages, there are several forms of labor arrangements, depending on the contents of work, availability of workers, and the relationship between the hirers and the hired (Table 5.1).

The first category, kindred labor, is the utilization of labor of one's close relatives, who may happen to visit and stay for a while, helping the host in various farming activities. This kind of cooperation is often made on the spot by unexpected guests and is thus not always available. Although there is no clear reward in this type of labor, bush meat or plantains may be offered as a gift when the guests return to their home village. The second category is *société*, a French word meaning "association". This is a form of reciprocal labor exchange in which villagers share labor in rotation during the clearing and weeding of swidden fields as well as in weeding and harvesting the cacao fields. Should one party fail to reciprocate there is no feeling of guilt, as is typically the case where a calculating type of reciprocity is used, for example in Southeast Asia.

The third category *grouper*, also derived from a French word meaning "to gather," may best represent the festive nature of cooperation in Fang society. This method is utilized in the clearing of swidden fields and breaking cacao fruits. The work is done in one day and, after the work, meals and alcohol are provided as a token of gratitude. Unlike *société*, there is no need to return equivalent labor. The last category – employment – is divided into two sub-categories; *job* ("a part-time job") and *contrat* ("contract").

Table 5.1 Labor arrangements among the Fang

Types of labor		*Ethnic groups*	*Contents of work*		*Reward*
			Swidden cultivation	*Cacao production*	
Kindred labor		Fang	All	All	None
Société		Fang	Clearing, weeding	Weeding, harvesting, transporting	Same amount of labor, alcohol, tobacco
Grouper		Fang	Clearing, weeding	Breaking fruits	Meal, alcohol
"Employment"	*Job* (day work)	Baka, Fang, "Anglophones"*	Seeding, harvesting	Weeding, harvesting, transporting	400–500 CFAF,** farm products, alcohol
	Job (area-based)		Clearing, weeding	Weeding	10,000 CFAF (for clearing 1 ha), 1,000–2,000 CFAF (weeding for 0.5 ha), farm products, alcohol
	contrat		None	All	10,000–50,000 CFAF (for 3 months), farm products, alcohol

Source: Field survey.

Notes

* "Anglophones" stands for migrant laborers hailing from anglophone areas in north-western Cameroon.

** €1 was equivalent to CFAF655.957 (CFA franc) in 2007.

Job is further divided into (1) one-day work, and (2) work based on area (e.g., weeding for 0.5 hectare), both with a reward of cash or equivalent alcoholic drinks (or tobacco). *Contrat* is a long-term contract of labor used especially during the season of cacao harvesting.

Société and *grouper* are useful especially for those who have returned from urban areas and started to cultivate anew. Even without cash, they can secure a labor force by providing free alcohol and bush meat (in the case of *grouper*). Bush meat is the most favored meal. Popular alcoholic drinks used in a work party are either palm wine or spirits made from cassava and maize. All these foods and drinks are made from the produce of hunting, gathering, or farming.

Cooperative labor is an integral part of the harvest and post-harvest processing of cacao, which requires a large labor force for a very short time. During the cacao harvest season, which lasts approximately three months from September to December, villagers employ all these forms of labor arrangement except that of *société*. Especially in the breaking of cacao fruits, villagers always practice *grouper*, as this process has to be finished in one day, with the seeds taken out from the fruits to be wrapped in banana leaves on that same day, in order to facilitate fermentation. To finish this work in one day, friends and relatives in and outside the village are asked to come to help that endeavor. After finishing the work, alcoholic drinks and meals are provided.

The skills that villagers have as "generalist managers" are closely associated with cacao production. In preparation for the cacao harvest, most producers direct their efforts to seeking bush meat and palm wine in order to secure enough labor. A Fang man, closely observed in 2007 by Sakanashi, prepared palm wine in greater quantities than usual in order to offer in *grouper* work. He also secured an abundance of bush meat by gun-hunting himself or through asking the Baka to procure bush meat for him. Some Fang even borrow money to prepare spirits to demonstrate generosity. In the cacao harvest season, the sales of palm wine and bush meat are also boosted due to a greater circulation of cash. Thus, in contrast to the commonplace view that cash crop production eliminates outdated hunting activities, there is a mutually reinforcing relationship in the study area of southern Cameroon.

Symbiotic relationships with the Baka in cacao production

Shifting cultivators in the African rainforest have often established a symbiotic relationship with neighboring hunter-gatherers so they can exchange farm produce and other items for labor or bush meat. Such a relationship has commonly been characterized as a fictitious parent-child relationship or patron-client relationship (Turnbull 1961). The introduction of cash crops may have strengthened or transformed such a relationship. Some scholars argue that cash crop sales made cultivators richer, thereby widening the economic gap between peasants and hunter-gatherers (Wilkie and Curran 1993). Based on his research in southeastern Cameroon, Hayashi (2000) asserts that penetration of the cash economy, partly brought by the development of cacao production, transformed the relationship between cultivators and the neighboring Baka into a mere employer-employee relationship through the medium of cash.

The Fang cacao producers depend heavily on Baka labor, particularly during the season of cacao harvesting. As shown in Table 5.1, there are two ways to mobilize their labor; *job* and *contrat*. The one-day *job* is especially useful for smaller cacao producers, but there is a risk of "no-show"; a Baka may fail to come because of the need to attend to other businesses. To be sure that they will turn up, employers, in addition to offering a wage, must provide the Baka with incentives in the form of palm wine, spirits, tobacco, and farm products during and after work. Sometimes, the exchange, however, works the other way. When the Baka need cash (or farm products), they turn to the Fang asking for *job* or gift.

Because *job* is uncertain, many villagers (25 out of 38 cacao producers in 2007) make a more stable, long-term labor contract, *contrat*, which runs for two to three months of the cacao harvesting season. There are two kinds of *contrat*. In one form, the Baka come to work from their home villages each day. The other is a live-in system, by which they stay at the home of a Fang villager for several months, not only working to harvest cacao but also to help the host in hunting and farming activities, even with domestic chores including drawing water. This live-in *contrat* is a good deal for the Baka who in addition to their wages are provided with daily meals as well as alcohol and tobacco more frequently than the practice is in the single-day *job*. Some Baka come to stay and bring along their family members (wife, children, brothers and sisters, in-laws), who may also work for the host. The host Fang, for his part, behaves as a generous patron to the entire Baka family and is often referred to as "my father" or "my brother" by the Baka.

This, however, is not necessarily a fixed patron-client relationship and the Fang often change their Baka partner. At the same time, there are some cases in which a Fang and a Baka retain a stable, long-term relationship, calling each other friends. The Fang may evaluate the Baka from the viewpoint of diligence, and the Baka, on their part, may attach great importance to the generosity of their patron. The Fang patron, therefore, must build a good relationship with a hard-working and good-natured Baka by continuously making a gift of alcohol and tobacco, a social relationship that is not just like managing a regular employer-employee relation.

Another feature characterizing *contrat* is that unlike contracts in capitalist society the wage is not predetermined. It is determined and paid at the end of harvest season, after the Fang have sold the harvested cacao. The amount is set by the host, depending on the price of cacao, the amount of debt the Baka owes him, and the closeness of the relationship between the host and the Baka. Upon receiving cash, the Baka workers tend to squander their hard-earned wages on clothes, tobaccos, and alcohol, displaying their penchant for "immediate return" (Woodburn 1982: 432). Purchased liquors and tobacco are immediately consumed together with other Baka and the Fang. Thus, in a sense, the existing economy of sharing is activated by cash flowing into the village during the cacao harvest season.

"Green" policies and peasant livelihoods

Beginning in the 1990s, under pressure from international aid agencies, the Cameroonian Government launched a series of "green" policies for the conservation of rainforests. As part of its Structural Adjustment Programs, which promoted market liberalization in developing countries, the World Bank required each government to establish laws to protect the environment in exchange for providing loans. Environmentally friendly policies and laws were also welcomed by the UN and other donor agencies (Goldman 2005). In the case of Cameroon, such green policies included a restriction on timber exports and the establishment of national parks and community forest areas. For example, the 1994 Forest Act requires the implementation of zoning, dividing the rainforest area into separate forest reserves, agricultural and residential areas, and community forest areas, along with the areas assigned for commercial logging. At the same time, the government in collaboration with international donor agencies embarked on the restriction of hunting activities in the rainforest. These hunting regulations include a ban on hunting specific animal species and a restriction on materials used in making wire traps, ruling out the popular metal wires. Gun-hunting now also requires a special permit. Violation of this law can lead to imprisonment and/or a heavy fine. Recently, the authorities have clamped down more severely on illegal hunting, sometimes leading to a gross abuse of power by the security forces. Furthermore, enforcement is very uneven and ineffective. In the study area, for example, there is a desktop plan for zoning the forest, but it differs widely from villagers' actual land use patterns (Sakanashi 2011). For example, the hunting field of the villagers covers an extensive area well beyond the boundaries of the community forest or agricultural areas identified in the zoning plan.

Cacao agroforestry has also been promoted by international donors as a way of combining cash crop production and conservation of the forest and its biodiversity.[5] Southern Cameroon has notably drawn attention to itself as an ideal place for establishing a model to promote cacao-based agroforestry, since villagers have a long tradition of practicing an indigenous form of agroforestry, leaving shade trees and growing other useful trees in the cacao fields (Sonwa et al. 2007).

The assumption that a more scientific and rational mode of agroforestry can be built on the indigenous foundation may prove overoptimistic.

First, as Shikata (2013) rightly points out, it ignores the fact that present-day cacao production is well integrated into the existing system of food production. As we have seen, the introduction of cacao as a cash crop has never put an end to swidden agriculture which makes sustainable food production possible. Second, villagers do not always depend on cacao production as the only source of income. Cacao fields may be abandoned at any time, due to a shortage of labor (particularly due to an exodus of youth to urban life and jobs); alternative sources of cash income (including hunting, non-timber forest products, or other cash crops such as orange, pineapple, and tomato); market price fluctuation; and so forth. Thus, villagers do not always stick to cacao production, attaching more importance to food production, whereas the state and foreign donors wish to promote permanent agroforestry to achieve an increase in production as well as forest conservation.[6]

Behind the promotion of agroforestry, as well as the restriction on hunting may lie an implicit assumption that villagers should be engaged solely in sedentary farming including cacao production. We have demonstrated, however, that hunting and gathering activities are indispensable for cacao production to secure the necessary labor force during the peak season. If their hunting and gathering activities are restricted more severely, cacao production would no longer be feasible. Besides, in periods of severe cash shortage (such as the post-harvest season locally called "the season of death"), villagers are heavily dependent on the sales of bush meat and non-timber forest products to pay medical bills and tuition fees. Thus, "green policies" promoted by the government and international donors are a potential threat to the survival of villagers, like the Fang and the Baka, who utilize natural resources in a multifaceted way.

Conclusions

This chapter has shown how Natural Society is still very much present in the way shifting cultivators in southern Cameroon and other wooded areas organize their livelihoods. They subsist on a form of swidden agriculture that is basically dependent on the nutritional cycle of nature to restore soil fertility. At the same time, they rely on hunting and gathering, making them focus more on managing several livelihood activities than on specializing in improving land use for higher productivity. Their grounding in Natural Society is further illustrated by their readiness to consider consumption needs that go beyond single households. It is a principle of honor for the haves to share with the have-nots because in a livelihood system where consumption, not production, is the ultimate objective, failure to engage in communal sharing would be a violation of the basic tenet of the political economy.

The introduction of cash crops is normally considered to bring a drastic change in such a "traditional" pattern of livelihood, notably by replacing fallow-mixed cropping with a system of monoculture on a permanent field. Such a move would typically also lead to greater reliance on a cash economy. As we have seen in this chapter, this has not been the dominant trajectory among the Fang. The introduction of cacao has not necessarily eliminated Natural Society elements such as multi-subsistence, the norm of sharing, and investment in interpersonal relationships. Cacao production has been incorporated into the existing system of swidden cultivation, and the labor-demanding nature of the cacao harvest has led to enhanced hunting and gathering activities, the products of which are invested in maintaining social relationships with neighboring hunter-gatherers and peasants. The process of production is inseparable from the consumption process particularly in the event of cacao harvesting, of which feasting participants are an indispensable part. Producing cash crops is only one option among various livelihood activities for local people and does not serve as the only lifeline as the case tends to be in Agrarian Societies.

A greater threat to the livelihoods of people like the Fang and the Baka are the policies of the Cameroonian Government, boosted by international donors, notably, the idea that the rain forest should be zoned and divided among specific categories of users. This arbitrary division of the territory on which the Fang villagers and their Baka neighbors depend for their livelihood could, if fully enforced, lead to the proletarianization of these people and the end of a livelihood system that has worked well for generations. The only silver lining in this scenario may be that official plans and policies rarely get implemented as intended. Like the forest zoning plan discussed above, it may remain a desk product and Natural Society, therefore, will continue to be part of the livelihood systems of southern Cameroon for still some time to come.

Notes

1 "Minor subsistence" is a term coined by Japanese anthropologist Takeshi Matsui, referring to hunting and gathering activities that are not significant in the sustenance of people's lives in terms of caloric intake, but are continually practiced from generation to generation to add materials which may be indispensable in local culinary culture (Matsui 2004).

2 The following accounts of the Fang are based on the first author's comprehensive fieldwork carried out (on and off) between 2005 and 2008.

3 The study villages belong to the Commune of Mintom, Dja-et-Lobo Department, located in the eastern part of South Province. The population density as of 2000 in Dja-et-Lobo Department is 8.64 per km^2 as compared to 11.02 per km^2 for South Province, 35.26 per km^2 for Central Province, and 47.54 per km^2 for South-West Province (Ministry of Economy and Finances, Republic of Cameroon 2000).

4 Cacao fields are also used as hunting grounds and hunting camps by the Fang.

5 The agroforestry system is not only about forest conservation and effective land use, but also about the social integration of indigenous people, who are otherwise not willing to settle in one place and are therefore difficult for the government to control (Combe 1982).

6 It is not to say that cacao-based agroforestry always falls short of our expectations. Establishing cacao agroforestry may be a possible option in progressive cacao producing areas in Ghana and the Ivory Coast, where improved varieties of cacao have been actively introduced by migrants. These modern varieties are resistant to sunlight and do not need shade trees, thereby contributing to a rapid deforestation in these areas (Ruf 2011).

6 Unique features of African agro-pastoralism

Adapting life and sharing wealth in a fluid environment

Yuichiro Fujioka, Yudai Ishimoto and Tadasu Tsuruta

Introduction

Agro-pastoralism is a unique form of livelihood which can be found across the semi-arid parts of the African continent. In his seminal work, Murdock (1959: 18) classified subsistence economies in rural Africa into five different categories: (1) hunting and gathering; (2) fishing; (3) primarily agriculture; (4) primarily pastoral; and (5) agricultural and pastoral (see Figure 1.2, p. 15). The latter is one of the most widespread ways of living, particularly in the ecological transition zones between rain forest and desert. In this environment dominated by savanna vegetation, both rain-fed agriculture and animal husbandry are practicable, thereby fostering a mixed agro-pastoral economy. In contrast to mixed-farming in Europe, where farming and livestock husbandry are closely linked to each other, agro-pastoralism in Africa has usually been viewed as lacking an organic relationship between the two.[1] While mixed farming in Europe focuses on the increase of agricultural productivity and intensive land utilization, African agro-pastoralism puts the emphasis on multiplying livestock rather than improved farming. This livelihood pattern is primarily an adaptation to precarious climatic conditions. To survive recurrent drought or for other reasons, agro-pastoralists have frequently relocated and accommodated themselves to new circumstances by combining farming, livestock husbandry, and even hunting and gathering. Unlike the shifting cultivators in the forest, however, those on the savanna accumulate wealth in the form of cattle which due to disparity in ownership leads to a measure of social stratification.

Drawing on intensive empirical research in Burkina Faso and Namibia, this chapter focuses on agro-pastoralism as another case of non-agrarian features of rural livelihoods in Africa. It begins by discussing the most important features of African agro-pastoralism that have been identified through ethnographic evidence gathered primarily from East Africa. This brief general discussion is followed by two case studies of contemporary agro-pastoral communities in somewhat contrasting environments: the Kel

Tamasheq in northeastern Burkina Faso as a case of people living in extreme environmental conditions, and the Owambo in north-central Namibia, as a group of agro-pastoralists living in somewhat better yet harsh conditions. Both cases highlight the importance of diversifying livelihoods and sharing foodstuffs as part of adapting to changing socio-economic circumstances.

The unique features of agro-pastoralism

Agro-pastoralists are distributed mainly in the semi-arid zones across Africa, which are also known as areas prone to drought and famine, along with other natural calamities including animal diseases (such as rinderpest) and locust outbreaks. In these conditions, agriculture is problematic therefore livestock often become an insurance against crop failure.

Jiro Tanaka (1982) argued that pastoralists and hunter-gatherers are better adapted to an unstable natural condition than sedentary cultivators as they are more mobile and thus have access to lands with more diverse flora and fauna. Agro-pastoralists, however, have developed their own strategies to cope with such threats as drought. When it occurs, they may rearrange grazing routes and watering places (Gallais 1977), utilize various wild plants and animals to survive (Bernus 1980; Ahmed et al. 2002), or replant crops in unutilized lowlands (Watts 1983). A study of the Gogo agro-pastoralists in central Tanzania showed how they constantly relocate their homesteads in search of pasture and water for their livestock. The idea of owning clearly demarcated plots for cultivation makes no sense in such conditions (Rigby 1969: 26–43).

Their mobility is made possible by changing the balance between livestock husbandry and agriculture, which must be managed by ecological zone and individual household. According to Katsuyoshi Fukui (1969a), who carried out field research among the agro-pastoral Iraqw in northern Tanzania, households in the grasslands put more emphasis on agriculture than keeping livestock, while those in wooded savanna and swamp areas tended to concentrate more on rearing livestock than on cultivation. Such malleability can be attributed to the fact that cultivation and herding are not necessarily linked to each other in the household economy. Given limited family labor, the more one concentrates on herding, the less farmland can be cultivated, and vice versa. In other words, in the African context, agriculture is inversely related to pastoralism (ibid.). To deal with this, foods are exchanged between people living in different ecological conditions. These types of institutionalized exchange networks – *inaoha* – are created not only between keepers of cattle and cultivators but also between cultivators specializing in different crops (Loiske 2004: 110–11). Summarizing the overall role of agriculture and pastoralism in the Iraqw subsistence economy, Fukui (1969b: 63, 70–5) concludes that, for the Iraqw, the value of agriculture lies in supplying food, while the value of a herd stems from its role in keeping and accumulating property. Livestock products (milk and meat) are less important as a source of food than agricultural products. Livestock is important as a symbol of wealth and a source of prestige. At the same time, it can also function as a savings bank, as both cattle and small stock can be exchanged for other goods. Generally, villagers are more interested in increasing the number of livestock than in expanding areas under cultivation. Cash income from farms tends to be used in purchasing livestock to boost savings.

Livestock also play a crucial role in the reproduction of societies particularly through the transaction of bride-wealth. Peter Rigby's classic work on the Gogo (Rigby 1969) describes in detail how their kinship system functioned through the sharing and exchange of cattle. Although their subsistence depends primarily on agriculture, their value system attaches prime importance to cattle. Livestock, particularly cattle, are the most important form of property, and families and relatives are connected intricately through the rights and obligations attached to cattle. The firmest ties are established through the exchange of livestock as bride-wealth for women, thereby creating new conjugal units and complexes of affinal relationships, which in the next generation become closer cognatic ties.

Unlike traditional communities of shifting cultivators, there is a certain kind of social stratification based on cattle ownership in agro-pastoral societies but like in the case of shifting cultivators in the forest there is a "levelling mechanism" by which those wealthy in cattle have an important role in distributing necessary foodstuffs among their resource-poor neighbors and relatives. Cattle trusteeship, which is found widely in agro-pastoralist and pastoralist societies, can give poorer villagers access to milk and manure (Rigby 1969: 43–53; Loiske 2004: 110–11). Whatever social stratification that develops in these societies, however, is not permanent, because livestock can perish anytime either because of such disasters as drought and diseases or be exhausted for social and economic reasons including the payment of bride-wealth and the sale of livestock to purchase grain. Even if the number of animals is retained until the owner's death, the fortune is often dissipated by his heirs. Thus, disparity in livestock ownership itself does not normally lead to the establishment of fixed class relations, unlike what happens in Agrarian Societies, where social relations are stratified into a class of landowners and another of impoverished tenants or sharecroppers. It is with these observations in mind that we now turn to the two case studies.

The Kel Tamasheq agro-pastoralists of northeastern Burkina Faso[2]

The Kel Tamasheq, also known as the Tuareg people, is an ethnic group living on the southern periphery of the Sahara Desert. Speakers of the Tamasheq language, one of the Afro-Asiatic languages, are widely dispersed over several different countries, namely Algeria, Libya, Burkina Faso, Mali, and Niger. In this section, we focus on a Tamasheq village in northeastern Burkina Faso, located in Markoye Department, which shares a border with Niger and Mali. The Tamasheq people are a typical marginalized ethnic minority in Burkina Faso, making up only 1 percent of the country's total population (14 million as of 2006) (Ouedraogo and Ripama 2009).

The livelihood of the Tamasheq is determined by an extremely harsh and unstable climate and other physical conditions of the semi-arid Sahel. A year is clearly divided into rainy (June to September) and dry (October to May) seasons. In Markoye Department the average annual rainfall between 1960 and 2004 has been estimated at 350 mm, with the highest record (565 mm) in 1964, and the lowest record being 184 mm in 1987. In addition to insufficient and unstable rainfall, outbreaks of desert locust storms sometimes occur throughout the Sahel. For example, in 2004, such a plague raged across northern and western African countries, swarming into the study village as many as seven times between September and October. Agriculture and pastoralism are the main livelihood activities, supplemented by occasional gathering activities of wild plants. Recently, most young men in the village have shown an increasing interest in moving to work in urban areas.

In 2005 the study village had a population of 146 divided into 32 households, mostly members of the Oudalan clan who are considered descendants from a common ancestor. Although previous studies (e.g., Ericksen 2010) reported that Tamasheq society is a class society made up of nobles, soldiers, and slaves, such a hierarchy was not observed in the study area. The daily social relationship among villagers is based on their kinship with families related by blood and marriage helping each other. Key social units in the study village are the homestead and the household. A homestead is a cluster of houses and is composed of one or many households. Twelve homesteads were found in 2005, scattered throughout several locations in the village. Residents in a homestead form a basic unit of production and consumption. A household consists of either a nuclear family or a polygamous family. Normally, the heads of households in a large homestead are closely related to each other, being parent and child, brothers and sisters, or cousins.

Livelihood activities

Agricultural production provides basic subsistence for the village residents, however unstable it may be. Their primary crop is pearl millet, supplemented by sorghum, cowpea, peanuts, and roselle. Main farming activities consist of: (1) clearing fields immediately before the rainy season begins; (2) sowing seeds after the first heavy rain; (3) weeding during the rainy season; and (4) harvesting immediately after the rainy season is over. Farm labor is drawn from the members of households in the same homestead.

Members give priority to their own household field but once finished they continue work by helping another household in the same homestead. This practice is especially common during times when a lot of manpower is needed in a short period such as when sowing seeds.

Livestock keeping is another important livelihood activity. Major domestic animals in the study area are cattle, goats, and sheep, along with donkeys and a small number of dromedaries. Livestock are kept for the purpose of sale, milking, and as an insurance against crop failure. Although a small stock is kept by all homesteads, the number of cattle differs considerably according to each homestead, ranging from 63 to none. Cattle, goats, and sheep are also used as bride-wealth. The standard price of bride-wealth is either three heads of cattle or 30 small stock, which are not easy to furnish for a man of marriageable age who normally must purchase animals or solicit close relatives for a gift of animals. Livestock is managed collectively by members of the same homestead. To save herding labor, the work of grazing is rotated among male members in the homestead.

There is a complementary relationship between agriculture and pastoralism. Farming activities supply crop residues to feed animals. Livestock, in turn, provide dung and urine for soil fertility improvement (Powell et al. 1998). In 2002 and 2003, when drought hit the area, villagers tried to make maximum use of animal dung and crop residues. They moved their houses and kraals to the field after crop harvest so that animals could stay in the field for a longer time than usual, eating crop residues and dropping dung over the field. Dung in the kraal was also spread over the field every three months, and the location of the kraal itself was shifted to change the location of dung accumulation. At the same time, crop residues were collected to prevent them from being consumed by members of other homesteads and villages. Thus, at times of emergency, there is enough flexibility to move the whole kraal and integrate livestock keeping and cultivation in ways that resemble the mixed farming system in Europe. Crop residues, however, were not enough to feed their stock in the dry season, so they had to be grazed in more distant places than in normal years.

Gathering wild edible plants is still an important livelihood activity, especially in years of crop failure. Among a variety of plants used for food, the most important ones in terms of caloric intake are grains belonging to the *Poaceae* family and tubers of water lilies (*Nymphaea* sp.), both of which are also used as famine foods. As the growing period of poaceous grasses is shorter than cultivated grains, their seeds can be collected during the pre-harvest period when food shortage may be acute. This gathering activity is carried out in swampy areas along seasonal rivers for a period of one month. There are two ways to gather the grains of poaceous grasses; (1) collecting matured seeds by basket before they fall from their ears, and (2) gathering fallen seeds by brooms, and then sifting them from sand. The former method is more effective than the latter, because it requires only a few moves and can be completed in a short time by a small number of people. Villagers mainly use the collection by basket in normal years and resort to collecting fallen seeds primarily in years of poor harvest, an exercise that involves all family members in full force. In past years of serious crop failure villagers moved temporarily to an area abundant with edible poaceous grasses, staying there and eating mainly their seeds for as long as several months. This practice has declined considerably these days, but water lilies continue to be important, especially as famine food, because their tubers can be collected all through the dry season when the water level is low. These tubers vary in taste, some being quite bitter and harsh, but in years of poor crop harvest, villagers will eat even the less appetizing varieties.

Recent transformation of the livelihood system

The livelihood system of village members has been transformed considerably due largely to the changes in ecological and socioeconomic conditions in the past half century. Before the 1970s, most homesteads owned a considerable number of cattle and small stock, which were sold to purchase foods in years of poor harvest. To graze a large herd, transhumance was necessary whereby herders and livestock moved seasonally some 40 to 50 km away from the village to areas where water and fodder plants were available.

In the 1970s and 1980s, serious droughts occurred extensively in the Sahel, causing catastrophic damage to pastoralism in the area. Droughts in 1972 and 1984 were particularly devastating in the study area.[3] Because the growth of pasture was very poor in those years, villagers experienced massive loss of livestock. It was difficult to recover the herd size in a short period, and villagers could no longer depend on income from livestock sales to purchase foods. As the total number of animals considerably diminished the demand for water and forage, the practice of transhumance was replaced by day-grazing within the environs of the village.

The most important socioeconomic change in recent decades has been the increased reliance on cash income from jobs in urban areas as an alternative to the sale of livestock out of shrinking herds. Such income has become especially significant at times of food shortage. The income from migrant workers became the most important source for purchasing foods during periods of food shortage. The more successful migrants spent their surplus cash not only to buy daily necessities including foods but also to increase the size of their herd. Newly bought animals either became a part of the owner's own herd or were mixed with herds of relatives in various places in order to conceal the increase in assets. This study shows that by purchasing livestock with income from urban work one homestead got its herd size back to the pre-1970s level or more. In this respect, labor migration has contributed to restoring pastoralism in the area.

Redistribution of foods in the consumption process

Despite recurrent droughts and times of food shortage, villagers in the study area have been excluded from the public social security system (such as emergency food aid), because of their marginalized position as a minority ethnic group living in remote areas distant from the capital city. Although food aid is occasionally distributed by the Burkinabe government, foreign donors, or non-governmental organizations (NGOs), it is unreliable and unpredictable, the amount is insufficient, or may not be provided at an appropriate time. Villagers, therefore, resort to self-help measures to cope with food shortage, including purchasing foods and collecting wild plants. Figure 6.1 shows the breakdown of caloric intake from starchy foods of two homesteads. While Homestead X has some male members working in urban areas, Homestead Y has no such major cash earner. This figure reveals that, in the years of food shortage (2002/03 and 2004/05), villagers were primarily dependent on the purchase of foods and support from other homesteads along with wild sources of foods. The cost of foods is normally covered by the sales of livestock or cash income derived from urban work. Even in a year of normal rainfall (2003/04, 414 mm per annum), Homestead X relied considerably on purchased food. According to the data, between September and December 2004, 62 percent of the cash used for food purchase in Homestead X came from wage labor, while that of Homestead Y was totally covered by the sales of livestock. Interestingly, 28 percent of the budget of Homestead X came from the financial support (as cash) of wealthier relatives, while Homestead Y registered a pair of one-year-old sheep as gifts during that period.

At the household level, self-help solutions such as food sharing or group eating within one homestead are important, particularly at times of acute food shortage. According to this practice, members of a household

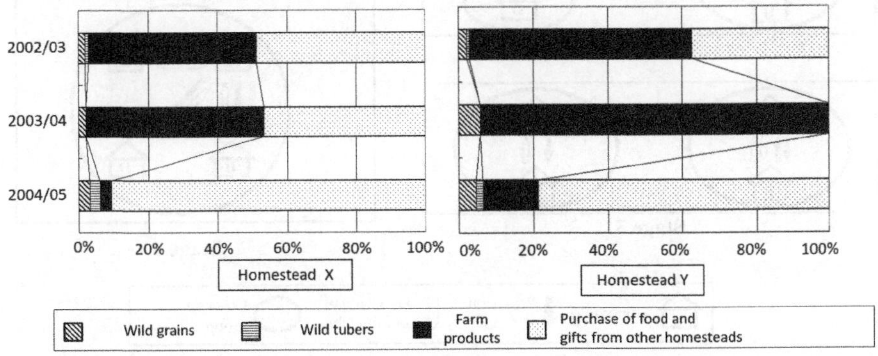

Figure 6.1 Estimated caloric intake from starchy staples.

Source: Field survey.

who have harvested an insufficient amount of cereals and exhausted them in the early stage, can gain access to food by joining the meals of other households with surplus grain. The case of the large Homestead X made up of six households is an interesting case study. Even in a year of normal harvest (2005/6), active food sharing was practiced in this homestead. To simplify a complicated process of co-eating, we focus on only four of its six households here (Households A, B, C and D). Household heads (or their spouses) are closely related to each other, for example as cousins or siblings. Household A has the largest number of adult males, thereby holding the largest acreage of land under cultivation and the largest number of livestock animals along with the largest cash income from urban work. The oldest woman in the homestead belongs to Household A. Household B is made up of a couple in their forties, along with the man's younger brother who was away from home, working at a distance. Households C and D are both made up of a young couple (a male in his thirties and a female in her twenties). There is a considerable age gap between women in Household A-B and those in Household C-D.

The process of food sharing in the four sample households between September 2005 and June 2006 can be divided into five sequential stages (Figure 6.2). During the harvest season (September to October), each household cooked and consumed foods separately, using materials

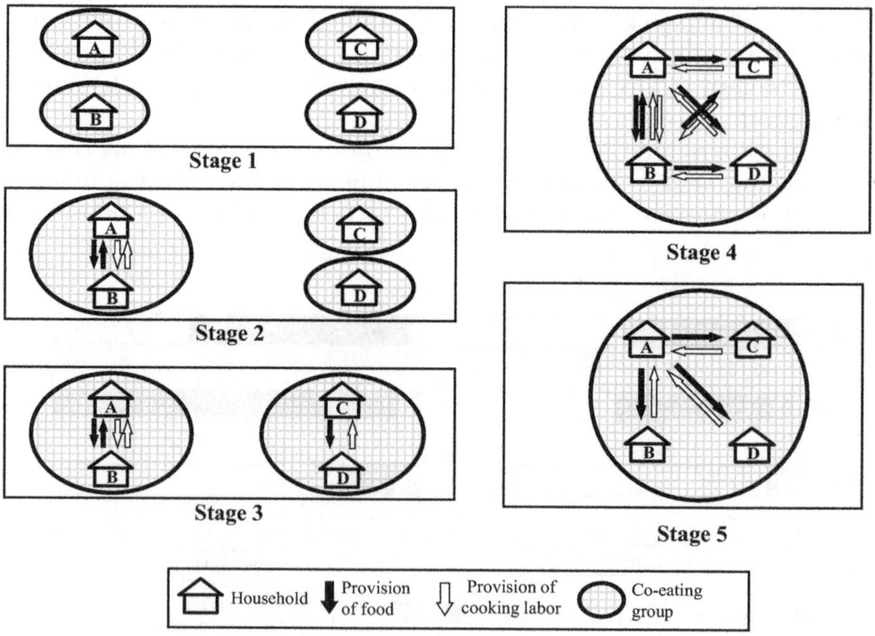

Figure 6.2 Successive stages of emergency co-eating.

Source: Field survey.

obtained from its own field (Stage 1). In the post-harvest season, all the adult men in the homestead gathered and ate together all the time. Therefore, the following description illustrates co-eating membership of female members (along with their children) of Homestead X, who also performed all the labor of cooking. Immediately after harvest, female members of Households A and B started to cook and consume foods together, while Households C and D continued to consume foods independently (Stage 2). As early as mid-November, the grain stock of Household D was exhausted, and its members then joined the meals of Household C. During this period, Household C provided foodstuffs and Household D did the cooking (Stage 3). After one week, Household C finished its grain and both Households C and D joined the meals of Households A and B. Members of Households C and D provided cooking labor at this stage (Stage 4). Finally, when grain stocks of Household B were exhausted in April, Household A became the sole source of providing foods, and cooking labor was provided by the other three households (Stage 5). In June, the grain stock of Household A had finally run out but by using income from urban work, additional grain was purchased and continued to be consumed by all four households.

This form of emergency food sharing is based on the everyday norm of reciprocity, which also manifests itself in a variety of ways in normal circumstances. For example, Households C and D tended to deliver a small portion of foods as a gift (locally called *takku)* to Household A, even before emergency co-eating started. Members of Households C and D also often helped those in Household A in their construction of houses and clearing fields. Such a daily supply of labor and food, in turn make it easier to ask for help from Household A at times of dearth. The point is that at the microlevel, mutual aid between households is not fixed and may be reorganized depending on circumstances. Is this also the pattern found among agro-pastoralists living under less adverse conditions? This is the subject of next section.

Wealth accumulation and sharing among the Owambo[4]

The Owambo is a Bantu ethnic group, living mainly in north-central Namibia and south-central Angola. Their territory spreads over a vast, flat plain (Cuvelai Basin) stretching from southern Angola to northern Namibia. A countless number of seasonal rivers form a web of channels, which flow primarily in the direction north to south only during the rainy season. The rivers often flood from the Angolan side to the Etosha pan, a seasonal lake located in northern Namibia. The climate in the area is semi-arid with the average annual rainfall being around 500 mm, falling mostly from November to April. The precipitation fluctuates considerably year by year, ranging from 237 mm (1961/2) to 998 mm (2010/11).[5]

Serious droughts occurred in the years of acute shortage of rain (as was the case in 1972, 1981, 1989, 1992, 2007, 2015, and 2016); while in the years of abundant rainfall, villagers suffered from floods which occurred in 1976, 1977, 1995, 2004, 2008, 2009, and 2011 (Mendelsohn et al. 2013).

The basic social unit in the village where Fujioka did his research is the nuclear or extended family residing in the same household.[6] The mainstay of rural livelihood of the Owambo is farming and livestock husbandry. The principal crops are pearl millet, sorghum, maize, cowpeas, Bambara ground nuts, and squashes. Pearl millet (locally called *mahangu*) is the main ingredient of their staple food, *oshithima*, a hard porridge made of millet flour. They raise cattle and small stock (goats and sheep) around the village, as well as in "cattle posts (*ohambo*)," the grazing camps located dozens of kilometers away from the village. Not all villagers can secure a cattle post of their own due to the considerable cost entailed in its establishment. The average number of cattle owned by cattle post owners is 60.5, while that of other villagers is 8.2.

Farming and livestock husbandry are mutually linked to each other in several aspects, in both production and consumption processes. Along with the use of donkey or cattle plough,[7] a direct linkage is observed in grazing within crop fields after harvest ("stubble grazing"). Villagers identify two benefits of this type of grazing. First, the livestock can graze on the crop stems, which contain a variety of nutrients. Second, the animals provide manure for the soil. Most households spread the manure piled up on the kraal ground into the fields before ploughing. During the dry season some villagers even shift the location of the kraal after several days within the fields in order to distribute the livestock dung and urine as efficiently as possible over a wide area. Thus, agriculture and pastoralism play complementary roles. One villager put it this way:

> We need a lot of cattle if we want to get enough pearl millet, and we need plenty of pearl millet if we keep a lot of livestock.

Apart from farming and grazing, they are also engaged in fishing, hunting, and gathering of wild plants and insects. As shown in Figure 6.3, while more than half of the dishes of sample households in terms of frequency are made up of livestock products such as beef and milk, villagers also utilize various other natural resources available in the surrounding environment including the African sharp-tooth catfish (*Clarias gariepinus*) caught in seasonal rivers, wild edible plants grown in the farmlands (e.g., *Cleome gynandra*), wild birds, and edible caterpillars appearing on mopane trees (e.g., *Gonimbrasia belina*, familiarly known as mopane worm). Villagers also often utilize indigenous trees such as doum palm (*Hyphaene petersiana*) and marula (*Sclelocarya birrea*) in a multifaceted way as construction materials, fuel, fodder, and materials for making alcohol. Marula beer (*omagongo*), a fermented drink brewed from the juice of marula fruits

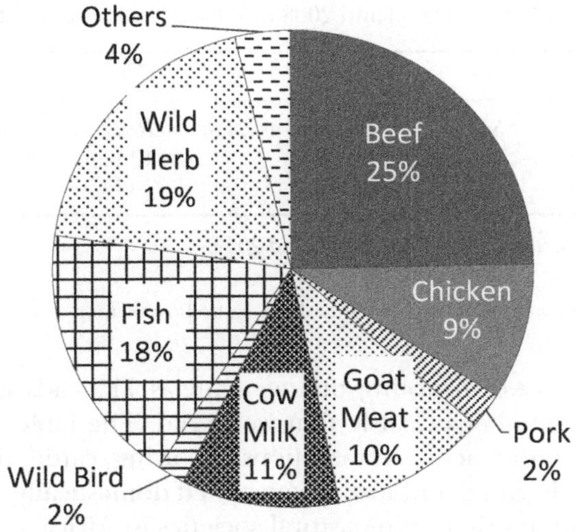

Figure 6.3 Frequency of dishes accompanying *oshithima*.
Source: Field surveys 2004 and 2005.

Note
The dishes in each meal were counted in four households from October 2004 to March 2005 (n=1047).

is considered the tastiest alcohol among the Owambo. From February to March, when marula fruits are available, women are busy squeezing the fruits and making the beer. In addition to self-consumption, marula beer is also actively gifted to other households, as further discussed below.

Since the study village is located near the town of Oshakati, the capital of the Oshana Region, some villagers earn wages from urban jobs and may own a concrete house and even a car. There is a considerable economic gap between wage earners and others, and economic disparity among households has grown recently. This gap is reflected not only in the number of cattle owned but also the size of the grain harvest, because households rich in livestock can utilize manure to enrich the soils in their farms. Whether rich or poor, however, households have one thing in common: they invest their income in farming and livestock keeping mainly for domestic consumption.

The socioeconomic significance of livestock

The Owambo use their livestock for several different purposes: food resources (meat and milk), animal power, wedding gifts, and exchangeable assets either for cash or for barter exchange between households. Table 6.1 shows that, between 2003 and 2008, a total of 78 heads of cattle were purchased by villagers, while only 40 were sold, suggesting a further

Table 6.1 Livestock transactions between 2003 and 2008 in Village U (30 households)

				(Unit: head)
	Bought	*Sold*	*Donated as a gift*	*Received as a gift*
Cattle	78 (43)	40 (29)	71 (23)	49 (13)
Goat and sheep	35 (14)	11 (1)	8 (2)	7 (1)

Sources: Field surveys 2002, 2003, 2004, 2005, 2006, 2007, and 2009.

Note
The numbers in parentheses indicate the transactions made by cattle-post owners.

accumulation of cattle. It is worth noting that as many as 71 heads were gifted, far exceeding the number of cattle that were sold. The table also indicates that there are more active transactions involving cattle than small stock, which are rarely gifted and mostly consumed domestically.

In Owambo society, like in other agro-pastoral societies in Africa, cattle have a special value as a symbol of wealth, playing important roles in cementing existing social relationships. Although it is uncommon for villagers in everyday life to give cattle to other households, steers are frequently given away at weddings.[8] Intimate relatives and friends of the new couple may donate a steer while others who are not so close to the couple, or cannot afford a steer, may give a castrated sheep, pots, baskets, or other tools used in daily life. The donated animal is slaughtered, and the cooked meat is provided to guests on the day of the wedding. Being mentioned by the master (or mistress) of the wedding ceremony as donor is an important occasion of displaying one's wealth.

Livestock is also important as an insurance against food shortage especially at times of climate disasters such as drought and flood. The Owambo usually store their millet in granaries called *eshisha*, each generally containing enough millet to supply a household for the whole year. They seldom sell the millet, which is stored in the *eshisha* for several years as precaution against crop failure. Grain stocks among households, however, differ considerably, with the poorer ones having to obtain additional millet using one of four ways: (1) purchasing (*oku-landa*), (2) donations from other households (*oku-gandja*) or two different types of barter exchange (3) *oshasha* and (4) *okupingakanitha*.

As shown in Figure 6.4, *okupingakanitha* is a bilateral exchange system, in which one person deals with another person, typically exchanging one full granary for livestock, especially cattle. *Oshasha* is the most important method. It is a multilateral barter exchange system involving several people who approach a person who needs millet (Figure 6.4). This person prepares something to barter for millet such as meat and fish, announcing that an *oshasha* event will be held on a specific day, whereupon villagers in need of meat or fish bring their millet and barter them according to the quantity of millet obtained. Among 54 millet transactions occurring in the

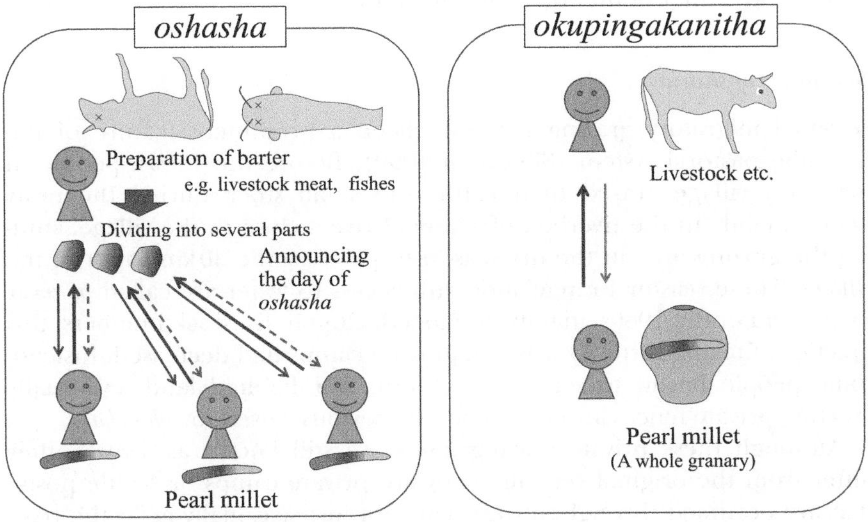

Figure 6.4 Two types of barter exchange.

Source: Field survey.

village between 2008 and 2012, *oshasha* accounted for 57 percent, exceeding purchase (32 percent, mainly from other villages), gift (11 percent), and *okupingakanitha* (2 percent). In the years of grain shortage, such transactions, especially *oshasha*, increase. Cattle or goat meat are not the only items exchanged at *oshasha* events. In 2011, for example, a household that brought two boxes of catfish eventually procured 83 kg of millet (equivalent to 51 days of food for an average household) from other villagers. Thus, this type of barter exchange, in which people regardless of economic status participate, increases the options to obtain grain not only for households with livestock, but also for poorer ones without it.

In Owambo society, people commonly recognize a certain norm of reciprocal help called *ethipa lyothingo* (literally meaning "neck bone").[9] Households are expected to help each other, for example, by offering daily labor or giving surplus foods, e.g., meat, milk and edible insects. This practice not only contributes to maintaining social relationships within and beyond the confines of the village, but also enlarges the opportunities for poorer households to obtain food. Wealthier villagers typically give meat or milk to those without livestock while the latter may make a return gift by brewing marula beer, the material for which is easily available in the surrounding environment. For example, from February to March 2007, 26 out of 30 households (87 percent) in the study village brewed marula beer, while 22 households gave the beer to other households as many as 525 times in total, 74 percent of which was for households in the same village (Fujioka 2010). The principal recipients of the marula beer were the wealthier households with livestock.

New forms of sharing

Seasonal migratory grazing has long been a prominent feature of the Owambo pastoral system (Siiskonen 1990). Before the 1980s, people in the study village grazed their cattle and small stock during the rainy season mainly in the riverbed of seasonal rivers close to the village, shifting the grazing area in the dry season to places some 30 km south of the village. These seasonal rangelands with access to water are called *ohambo.* By the end of the 1980s, mainly due to a decline in livestock numbers, this practice of taking cattle to seasonal grazing camps had decreased. Instead, some people began to settle on hitherto uninhabited land, eventually erecting private fences around the areas previously used for *ohambo.*

Although these private grazing lands are still known as *ohambo,* they differ from the original version. They are private camps or "cattle posts" that are occupied throughout the year, not just seasonally as in the past. Furthermore, *ohambo* land may now also be used to grow crops for consumption by the owner and his family. Herders are often employed as full-time workers in cattle posts. The grazing system has drastically changed due to the emergence of these permanent grazing camps. The livestock is kept in the cattle posts for most of the year (around September–May), although some cattle may be transferred to the village for several months in the dry season (around June–August) to graze in the crop fields. The cattle-post owners also collect resources not easily available in the study village including edible mopane worms, which are often brought back to the village as gifts for neighbors. Manure, which was formerly rarely gifted, is nowadays carried to the village by car and given to needy villagers.

Most of the cattle-post owners in the study village are affluent individuals who have jobs in town. The first cattle post by a villager was set up in 1982, and the number of cattle-post owners subsequently increased, particularly during the 1990s. By 2006, nine cattle posts had been established by eight households (27 percent of 30 households), all of which belong to the high-income stratum in the village (Fujioka 2010, 2014). To establish a cattle post, an individual must obtain permission from the headman of the area by paying a fee of N$600.[10] In addition, cattle-post owners invest a large sum of money to build fences and other facilities on the site. To manage their livestock, they employ local herders. The estimated cost for one year of operation is about N$11,100, almost equivalent to the price of seven female cattle, and thus representing an enormous expenditure for ordinary villagers.

At first glance, the sole goal of investment in cattle posts seems to have been to augment personal wealth but it is also an act of stabilizing production of grain for self-consumption at times of unstable weather conditions.

It is one link in a more complex chain whereby more livestock bring more manure to enrich the soil which in turn leads to the conversion of cash into shareable resources such as livestock, milk, manure, and edible insects caught at the cattle posts. These things are given to other households on various occasions, while cash is hardly used in transactions among villagers. Thus, the afore-mentioned norm of "neck bone" still guarantees that resource-poor villagers can get meat and milk from their wealthier neighbors in return for marula beer.

Conclusions

Three important conclusions can be drawn from the analysis of these two case studies. The first is that elements of Natural Society are very much evident in how agriculture and pastoralism complement each other in the livelihoods of the local people. Whether among the Kel Tamasheq in their harsh environment or the Owambo in their somewhat more benign circumstances, livelihood patterns are well adapted and designed to get the most out of what nature provides. The second conclusion is that even though a deterioration of the climate conditions has led to a greater reliance on cash income, the subsistence ethic continues to be relevant as more well-to-do households or homesteads share resources with the poorer ones, sometimes in terms of gifts, at other times as barter. Livestock is not merely a private asset; it is also a social wealth that must be shared among people either as bride-wealth (Kel Tamasheq) or a gift in the wedding ceremony (Owambo). The third conclusion is that policies that are based on the assumption that African societies are agrarian societies miss the main point about how livelihoods are designed and managed and how they can be further developed from within rather than being subject to ill-informed interventions in the name of modernizing agriculture. As the cases discussed in this chapter illustrate policies must consider what Mortimore and Adams (2001: 49) refer to as the "systematic understanding of indigenous adaptive capabilities," behind which may lie the generalist attitude of villagers toward their production system as well as their penchant for sharing foodstuffs to ensure that everyone has a chance to survive. The analysis of their production system, therefore, must be supplemented with an approach from the viewpoint of the mode of consumption, which highlights the non-agrarian features of their society.

Notes

1 Recently, as many agro-pastoral communities became sedentary, villagers have adopted some elements of mixed farming, especially the use of animal draft power and manure (McIntire et al. 1992).

2 This section is based on the fieldwork carried out by the second author between 2002 and 2008.

3 Annual precipitation in Markoye (approximate 20 km away from the study village) was 271 mm in 1972, 189 mm in 1984, as compared to 350 mm, the average between 1960 and 2004.

4 Data were obtained from fieldwork carried out by the first author in Village U located in Oshana Region on and off between 2004 and 2013 (33 months in total). The field research covered almost all of the households (around 30 households) residing within the village.

5 These figures are based on data in Ondangwa Town in northern Namibia, provided by Namibia Meteorological Service.

6 Formerly, the Owambo had generally lived in polygamous homesteads, each composed of several households. However, in the study village, which is located on the outskirts of the regional capital, homesteads have transformed themselves into separate households, all of which are made up of monogamous families.

7 Villagers formerly used to plough their fields with oxen, which have been replaced entirely by tractors in recent times. Only a few villagers plough with the aid of a donkey, because they cannot afford to hire a tractor.

8 Recently, bride-wealth is normally paid in cash, and not necessarily in cattle. Local rules about bride-wealth differ according to each subgroup. For example, among the Kwambi, one subgroup dominant in the study village, kinsmen of the bridegroom, in the past could have paid up to around ten heads of cattle as bride-wealth to the parents of the bride.

9 According to the villagers, this term bears a double meaning. One of the interpretations derives from the fact that, when people slaughter their livestock, they should give away livestock meat, which is hung around the necks of their neighbors. Second, it also connotes the norm that households should be strongly connected with each other, just as neck bones are tightly linked with each other.

10 One Namibian dollar (N$1) was equivalent to 0.13 US dollars (US$0.13) in 2004.

7 Food sharing among commercial rice growers

Persistence of the subsistence ethic in Kenya

Noriko Ito and Tadasu Tsuruta

Introduction

Placing peasants in organized settlements so that in such "uprooted" conditions they would become more open to productivity innovation is a recurrent theme in the history of agricultural policy in Africa. A "golden" opportunity to achieve this goal arose in the Kenyan highlands in the 1950s, when following the suppression of the Mau Mau Rebellion the British colonial administration embarked on a large-scale land improvement project to deal with African complaints about land shortage, a major reason for their poverty and political discontent. As part of resettling those who were detained during the uprising, as well as other landless villagers, the British established the Mwea Irrigation Scheme (MIS) with the dual objective of bringing peasants into the market economy and increasing the production of rice for Kenya's growing urban population by using irrigation.

Farming in the scheme has indeed been commercialized but it does not mean that "moral economic" practices have disappeared. Ito has confirmed through her research in the Mwea Scheme that the "subsistence ethic" is still very much alive in this commercialized setting. The chapter will elaborate on the livelihood patterns that have evolved in this environment placing it in contrast with the commercialization of peasant agriculture in Southeast Asia. Thus, it begins with a brief review of rice production in Southeast Asian rural communities before and after the Green Revolution. It is followed by a discussion of the social dynamics of sharing that continues to be at play in the Mwea settlement area.

Transformation in Southeast Asia

As this volume argues and demonstrates, governments and international donors have not understood that the reason why there has been no equivalent of the Asian Green Revolution in Africa is that its societies have never been agrarian. Because there is no tradition of specializing in certain crops, the interest of African peasant producers has been to diversify livelihood sources, placing their eggs, so to speak, in multiple baskets. Before the Green Revolution took hold in Southeast Asia in the 1960s, the peasantry was regarded as backward, similar to its counterpart in Africa. By the next decade, however, livelihood patterns had changed drastically. Rice, once regarded as a mere subsistence crop, had been converted into cultivation for cash (Tomosugi 1995: 78). Today, the extensive penetration of commercial farming has considerably weakened former rural cultures and communal bonds in Southeast Asia, to the point where one observer at the end of the twentieth century emphatically pronounced "the end of the peasantry" (Elson 1997). For instance, mutual aid customs among villagers such as labor exchange and rice-sharing practices have declined rapidly, evident in the rich Japanese scholarship on rural Southeast Asia that will be used in this chapter.

The situation before the Green Revolution

Most of the present Southeast Asian nations have their origins in the states established between the ninth and thirteenth centuries, based on sedentary rice-farming communities. Such a long tradition of Agrarian Society has resulted in the widespread monoculture of rice throughout the region, especially in lower basins and plains. By the late nineteenth century, immediately before the invasion of western colonial powers, there were indigenous states originating from the long-standing tradition of bureaucratic systems with rice farmers as a subordinate class. Furthermore, commercialization of rice occurred as early as the beginning of the twentieth century (Tsuruta 2011). This historical condition – a strong state and a penetrating market economy – constituted a favorable environment for realizing the "Green Revolution" some decades later.

Until the first half of the twentieth century, the subsistence ethic, or the notion that all members of the community should be guaranteed their minimal standard of living, had existed in the rural societies of Southeast Asia. Scott (1976: 5) argues that, along with the "safety-first" strategy in farming, a wide range of moral arrangements functioned to assure minimum subsistence needs for poorer villagers. In years of crop failure, peasants in distress would benefit from the right to use communal land, material support from patrons, reduction of tenant rates by landowners, and tax reduction or a supply of reserved rice by state.

Geertz (1963) showed how villagers in Java during the colonial period eked out a subsistence-level existence by "sharing poverty." By the 1960s, in the rice-growing villages of Java, landowners, tenants and landless villagers sustained a variety of cooperative arrangements regarding land tenancy and labor mobilization (Hayami and Kikuchi 1981: 155–7). For example, *derepan* refers to a form of labor utilization under which anyone can participate in harvesting and threshing for a specific farmer with participants receiving a certain share (locally called *bawon*) of the output.

A similar custom (locally called *hunusan*) existed in the Philippines (ibid.: 80–3). In yet another example, a traditional share-cropping system in Java called *maro*, the harvested rice was divided equally between the landowner and the tenant, on condition that the two parties also equally shared the cost of input materials, including seeds. This practice also functioned as sharing the risk of a poor harvest equally between them (Fujita 1990: 39–40).

Changes brought by the Green Revolution

The Green Revolution in Southeast Asian countries was carried out by each state, which constructed irrigation facilities before introducing the intensive seed-fertilizer technology in the 1960s and 1970s. In the irrigated areas, the agricultural extension service and public finance institutions helped disseminate high-yielding varieties of rice along with chemical fertilizers and tractors (Hayami and Kikuchi 1981). The increase in productivity of rice brought about three major changes in rice production and marketing: (1) enhanced commercialization; (2) new ways of sharing and redistribution; and (3) shifts in the landowner-tenant relationships.

In the irrigated areas overall, the ratio of rice for sale among the harvested rice of each household (hereinafter referred to as "selling rate" of rice) increased after the Green Revolution. In Central and East Java in the 1970s, 55 percent of surveyed farmers in irrigated villages sold rice, as opposed to 20 percent in unirrigated villages. The larger the farm, the greater the selling rate and profit, due to the fact that large-scale farmers normally sold a great deal of rice at a higher price to major buyers such as cooperatives and large rice mills, whereas small-scale farmers tended to sell small amounts of rice to middlemen or in local markets at a lower price (Kano 1979: 130–1, 1981: 109). In the 1980s, around 60 percent of farmers sold rice even in unirrigated villages in West Java (Fujita 1990: 11). In irrigated villages in East Java, the selling rate of rice for sample farmers exceeded 70 percent (Koike and Kitamura 1987: 28). In the 1970s in irrigated villages in central Thailand, virtually all the farmers sold 100 percent of the harvested rice, while in unirrigated villages in northeast Thailand, the rate remained much lower (48 percent) even in the 1990s (Tomosugi 1995: 78; Nakada 1996: 616). In Central Luzon in the Philippines, the average selling rate of rice was 60 to 80 percent in the 1960s, and the selling rate of larger farms exceeded that of smaller farms in the 1970s (Umehara 1992: 140, 225).

The systems of sharing the harvested rice have changed along with the commercialization of rice. In the 1970s and 1980s the *derepan* (Java) and *hunusan* (Luzon) systems were gradually replaced by one that limited participation in harvesting, allowing only those who offered free labor in transplanting and weeding, the activities that precede harvesting. This new practice was called *ceblocan* in Java and *gama* in Luzon. Both old and female villagers were excluded from *ceblocan*, and the share received by the laborers (*bawon*) declined from what it had been in the *derepan* system (Hayami and Kikuchi 1981: 87, 159). In the 1970s in irrigated areas in central Java, large-scale farmers started to use the *tebasan* system, in which they outsourced the harvesting to an outside trader, thereby excluding the neighboring households from participation in the harvest (Kano 1981: 112–17).

Landowner-tenant relationships also changed after the Green Revolution. The traditional share-cropping system was replaced by a system of fixed-rent tenancy in central Luzon in the 1970s and in Java in the 1980s (Fujita 1990: 39–40; Umehara 1992: 156–7). Although the fixed rent was set lower than that of share-cropping, the tenant was at greater risk in the case of a poor harvest because he or she had to pay the fixed amount regardless of the size of the harvest and in addition had to shoulder all the management costs. For example, in central Luzon, landowners or employer-farmers might go as far as annulling the contract with tenants and laborers if they failed to pay rent or pay back rice that they borrowed in the hunger season at a high rate of interest (Umehara 1992: 158–9, 214–15).

Thus, the state-led introduction of Green Revolution technology considerably transformed not only the rice production system but also social relationships, undermining the function of former rice-sharing and risk-spreading systems employed to guarantee the subsistence of the underprivileged. This change has expanded the economic gap between rich and poor villagers although commercialization has proceeded without causing overt conflicts between villagers as overall rural economic growth has partly eased the predicament of landless villagers and small-scale farmers through increases in real wages, along with increasing opportunities for off-farm income (Kano 1981: 139–47). Overall, the conclusion is that the Green Revolution has been a successful social transformation taking development to a higher level in Southeast Asia. This trajectory stands in contrast with what has happened in the state-led irrigation project in Mwea.

Commercial farming in Mwea[1]

The MIS, located about 100 km northeast of Nairobi, is the largest rice cultivation area in Kenya. Construction of the irrigation facilities began in 1954. By 1960, about 2,100 hectares of land had been brought under irrigation (Chambers and Moris 1973: 434). The original settlers were members of the Kikuyu ethnic group, many of whom were suffering from a shortage of farmland due mainly to population increase. The initial residents included those who were detained at the Mau Mau Rebellion (Ishii 2007: 128–31).

The first generation of settlers

The initial tenants settled with their respective nuclear family (*macii* in Kikuyu language), typically made up of the tenant (adult man) himself, his wife and small children. At the outset, four acres of paddy fields were leased to each tenant. The original settlers called their cultivation area *maeka*, a term created by attaching the Kikuyu prefix *ma-* to the English word "acre," celebrating the fact that they now got their own farmland (Ishii 2007: 131).[2] Today, the irrigated area in the scheme (around 6,200 hectares in total) is divided into five sections, with each consisting of ten to 20 groups known as "units." Each unit has about 100 hectares (250 acres), although the cultivation area of each household in the same unit may vary (Njeru et al. 2016: 302–3). Recently, records show that there are more than 6,000 households in total (JICA 2015: 15).

The National Irrigation Board (NIB) was established in 1966 as an administrative authority for the distribution of irrigation water and for controlling rice production within three national irrigation schemes, Mwea being one of them (Carruthers and Weir 1976: 302–3). The board's management style was authoritarian and strict, in adherence with the 1967 Irrigation Act. The settlers were expected to comply with all the instructions given by the scheme managers (Ishii 2007: 171). The NIB provided settlers with tractors and other input materials, including seeds and chemical fertilizers. The settlers had very little freedom of choice or control in farming, branding or distribution of harvested rice, even in livestock husbandry like grazing, dipping and inoculation (Kabutha and Mutero 2002: 129). Under the dictates of the NIB, the settlers were obliged to sell almost all the harvested rice to the NIB, which in turn paid the settlers only once a year approximately six months after the harvest. They were permitted to bring home less than 10 percent of their harvested rice, which typically was not enough to sustain their families. Thus, the settlers grew maize and beans around their paddy fields to make *githeri*, the Kikuyu traditional staple food. Rice was consumed mainly on special occasions such as Christmas holidays and was rarely shared between households. Mwea Rice Growers Multipurpose Cooperative Society (MRGM), which was established in 1964 to promote settlers' prosperity, provided them with loans – secured on the harvested rice – for purchasing input goods and covering the cost of living for the whole year (Ishii 2007: 170–1). Mutual aid practices such as labor exchange among households were rare in the rice production process on the *maeka* lands. During the busy season, settlers used family labor including their young children. As the children grew older and became independent, the settlers began to use loans from MRGM to pay wages for day laborers from nearby villages in order to keep the rice production schedule set by the NIB.

The strict rules and regulations generated resentment and hostility against the board among the settlers. Especially controversial was the rule under the terms of the Irrigation Act that only the eldest son of the original leaseholder would be allowed to continue staying in the scheme after turning 18 years of age. All other children would have to leave. This policy ran contrary to the traditional Kikuyu norm of fathers transferring parts of their land to each son at his marriage (Kenyatta 1978: 20–52). Despite the official policy, first-generation settlers began to informally subdivide the *maeka* among their sons mainly in response to expectations among this second generation of settlers that they had helped their fathers with farm work when they were young. Without responding, the fathers would be viewed as failing to fulfil their role according to Kikuyu custom. Although these subdivided plots were too small to earn a living, most second-generation settlers remained in the area, eventually leading to open confrontation with the NIB.

The revolt of the second generation

The feeling among second-generation settlers that NIB limited their right to use enough farmland within the scheme eventually blew up into the open. In 1998, some young settlers successfully seized rice being delivered to the NIB and brought it to the MRGM, leading to a heated confrontation between the settlers and the NIB. A politician, who was the son of a first-generation settler, took up their complaint and accused the state of exploiting the settlers. The latter and the MRGM won this political battle and were eventually able to take over all aspects of running the MIS and elected new unit leaders as their true representatives. Finally, in 1999, the government liberalized the rice production and distribution system in the scheme (Ishii 2007: 181–9).

The riot in 1998 also precipitated two notable changes: (1) expansion of farmland outside the scheme area; and (2) diversification of the marketing of rice. Second-generation settlers have since rapidly expanded their own paddy fields beyond the boundaries of the original scheme area, an act which had formerly been prohibited by the NIB. These newly cleared paddy fields are called *jua kali* (a term denoting informal sector economic activities in Kiswahili). The rice production method on the *jua kali* farm is considerably different from that applied on *maeka* land where it is controlled (in terms of water use and cropping schedule) by the NIB and characterized by frequent use of chemical inputs and wage labor at considerable expense. Instead of purchasing seeds and chemical fertilizer from the MRGM or in the market, *jua kali* growers recycle seeds from the rice they harvest and use manure obtained from their own livestock. As there are no irrigation facilities on *jua kali* lands, the second-generation settlers cooperate with each other to supply water through plastic pipes from a nearby river. This has caused water shortages in the paddy fields belonging to the irrigation scheme (Ishii 2007: 190–2). The result is that the NIB was invited back to ensure control of the technical side of the production in collaboration with the settlers (Mati et al. 2011: 147).

While the second-generation settlers use a smaller portion of the cultivation area than the first-generation settlers, they make the best use of their vast social networks in the rice production process. As most of the second-generation males were born and raised within the settlement and married women living in the area, they are surrounded by friends and in-laws. They cooperate with each other in various phases of rice production, sometimes providing helpers with meals or unhusked rice as a token of gratitude. To save the cost of production, some of them exchange input goods including rice seeds and manure. One of the second-generation settlers explained that the cooperation in *jua kali* was motivated by the spirit and culture of *harambee*, a Kikuyu word for "working together," now commonly used in Kenyan Kiswahili. Most of them have also ceased to help their fathers in the *maeka* area claiming they are too busy trying to make a living of their own.

After the riot in 1998, most settlers stopped selling their rice to the NIB and waiting for its lump sum payment. Instead they sold it to the MRGM or directly to the millers and other buyers in the local market rather than waiting for the lump sum from the NIB. This new flexibility allowed them to earn the cash needed for school tuition fees, renovating houses, repaying debts and paying wages for hired labor. According to a survey carried out from 1992 to 2002 in 43 sample households in MIS, the average rice selling rate to the NIB declined from 90 percent in 1992 to 0 percent after 1999, while that to MRGM increased from 0 percent in 1992 to 36 percent in 2002. At the same time, the percentage of rice brought home increased from 10 percent in 1992 to 64 percent in 2002 (Ishii 2007: 187–9). Ito's field study of 47 households found that the average rice selling rate decreased from 90 percent before liberalization to 85 percent in the 2011/12 season. There has also been a change in the rice varieties planted. In addition to the existing Shindano BW type, most settlers have begun to grow basmati 370, an aromatic rice variety introduced in the early 2000s as a more promising cash crop.[3] The BW variety is grown primarily for domestic consumption. According to the same fieldwork, all the 47 sample households planted basmati, and 36 (77 percent) also produced BW. The diversification in production that occurred after the 1998 liberalization proved to be significant for how the two generations of settlers relate to each other.

The social function of rice sharing practices

As indicated above, there is a sharp discrepancy in the socioeconomic status of the first and second generation of settlers. Table 7.1 summarizes the average rice production and consumption of the 47 sample households which belong to the same unit of the scheme, including 12 first-generation settlers (average age 73) and 35 second-generation settlers (average age 48), who were interviewed in 2012. Farm size, income per head and rice production of the first-generation settlers greatly exceed those of the second generation. In addition, the family size of the second generation is larger than that of the first generation. While the latter typically have a surplus of rice, the former cannot afford to spare the quantity needed for home consumption.

The manner of dealing with the harvested rice differs considerably between the two generations. A sizable amount of rice was donated by the first generation to the other households (as much as 14 percent of their production). On average, the first-generation settlers sold half the harvested amount of basmati (around 52 bags[4]) to private traders and rice mills immediately after harvest. After keeping five bags of BW for self-consumption, they donated on various occasions 19 bags of BW to other households, in addition to intermittently selling the remaining stock of basmati when they required cash. Among the rice donated, around 40 percent was given to the other sample households in the same unit. In contrast, the second generation donated only 8 percent of their total production. Fourteen second-generation settlers did not donate rice at all. Instead, they sold most of it immediately after harvest. The selling rate of rice for the second generation (88 percent) is higher than that of the first generation (81 percent). The former received on average 12 percent of produced rice from other households as a gift, which is considerably higher than that of the first generation (2 percent). With few exceptions, most of them consumed all the BW rice at home or gave it as donation. As 71 percent of them were lacking BW for home consumption, they had to purchase rice in local shops. Second-generation settlers are also more dependent on off-farm income to pay for school fees and medical bills for their families including their young children. Overall, the amount of rice produced by each household has a positive correlation with the proportion of rice donated to other sample households, and a negative correlation with its selling rate.

Table 7.1 also indicates that a considerable amount of rice was transferred from first- to second-generation settlers in the surveyed unit. There are two main occasions on which the first-generation settlers give rice to other households: (1) social gatherings and rituals, and (2) emergencies like food shortages. The first-generation settlers brought with them the Kikuyu custom of getting together called *mucemanio,* in which around 20 to 30 neighbors bring food materials to cook and eat together in something like a "potluck" event, primarily in the post-harvest season (from December to February). In contrast to the original *mucemanio* held in their home villages, in which only a few elders discussed various issues such as ceremonies and conflicts among villagers, the *mucemanio* in the settlement scheme is organized as a social occasion to promote friendship among neighbors. Especially wealthier (mostly first-generation) settlers tend to donate considerable amounts of BW rice to neighbors on these occasions. Second-generation settlers, on their part, use the *mucemanio* to complain about their plight, mentioning that they are unable to buy enough food or pay for the education of their children. In addition to the *mucemanio,* first-generation settlers also donate food materials to their ritual kindred[5] on

Table 7.1 Average rice production and consumption of households by generation *(2011/12 season)*

Generation		*First*	*Second*	*Total*
		(n = 12)	*(n = 35)*	*(n = 47)*
Number of household (HH) members		2.75	4.57	4.1
Income per head (KSh)*		82,279	47,015	56,018
(Proportion of income from rice)		(81%)	(70%)	(73%)
Farm size (acre)	*maeka*	4.17	1.24	1.99
	jua kali	0	0.56	0.42
	Total	4.17	1.8	2.41
Breakdown of acquired rice (kg/year)	Produced: A	12,631	5,093	7,017
	Purchased	2	35	26
	Received from other HHs as a gift: B	311	630	549
	Total	12,944	5,758	7,592
Breakdown of the uses of rice (kg/year)	Domestic consumption: C	565	791	733
	Sold: D	10,175	4,501	5,950
	Donated to other HHs.**E	1,807 (758)	412 (221)	768 (358)
	Saved as seeds for the following year	17	6	9
	Total	12,564	5,710	7,460
Receiving rate (B/A)		2%	12%	8%
Domestic consumption rate (C/A)		4%	16%	10%
Selling rate (D/A)		81%	88%	85%
Donation rate (E/A)		14%	8%	11%

Source: Field survey, 2012.

Notes

All rice quantities are given as paddy (unhusked rice) equivalent.

* US$1 was equivalent to KSh84.2 in the year of survey.

** This includes the donations made at ceremonial occasions in which several HHs gathered. The numbers in parentheses indicate the rice donated to the other sample HHs in the same unit (included in the total).

ceremonial occasions in church, offering supplies for the participants in the event, including the pastor.

Yet another occasion for rice donation is to help those who are short of food, typically a gift by fathers to their sons. In the 2011/12 season, around 60 percent of second-generation settlers visited their fathers to ask for rice two or three times in the pre-harvest months between September and December, when food is in scant supply. A son's family may appeal for help by saying "your grandchildren are hungry." Fathers warmly welcome their sons' families by entertaining them with a meal and giving them ten to 20 kg of rice, thereby playing the role of the generous and dignified family head, the ideal image of the father in Kikuyu tradition. Thus, in contrast to the rice production process, in which intergenerational cooperation is rarely observed, the two generations interact through the process of sharing rice, thereby mitigating the economic gap between them.

The different rice-sharing patterns in the two generations reflect not only their socioeconomic status but also their moral principle that wealthy individuals should help those who are poorer. Table 7.2 indicates the different patterns of donation by generation. In the 2011/12 season, approximately 60 percent of all rice donated by the first-generation settlers went to their children (26 percent) and ritual kindred (33 percent). Thus, the largest percentage of rice is redistributed to second-generation settlers, as most of their ritual kindred belong to these families. Meanwhile, the second-generation settlers made rice donations largely to their brothers (37 percent) and in-laws (19 percent). These figures reflect the fact that most second-generation settlers donate rice mainly as a "reward" for those who assisted with work on their *jua kali* farms. This practice rests on a different rationale than the one used by the first generation.

The persistent connection between subsistence and reciprocity

We have so far examined how villagers in MIS have changed the way of dealing with the harvested rice after liberalization by focusing on the increased amount of rice allocated for home consumption as well as for active food-sharing practices. In contrast, post-Green Revolution farmers in Southeast Asia moved undaunted in the direction of commercializing rice production, which considerably reduced the role of reciprocal relations among rural dwellers. As we have noted elsewhere (Tsuruta 2008), reciprocal relations were separated from the subsistence needs of villagers in Thailand and came to have only symbolic roles in the spheres of gift exchange on ceremonial occasions. At the same time, the nature of reciprocity gradually adapted, influenced by the logic of the market economy. By referring to the typology of reciprocity used by Marshal Sahlins (1972), Sugimura (2004) argues that reciprocal social exchange

Table 7.2 Recipient of rice donation in each household by generation

Generation		*First*	*Second*	*Total*
		(n=12)	*(n=35)*	*(n=47)*
Average amount of donation (kg/year)*		676	105	250
Recipient of donation (%)	Parents or children**	26	4	19
	Brothers	15	37	22
	In-laws	17	19	18
	Other relatives***	6	15	9
	Ritual kindred	33	13	27
	Friends inside village	1	0	0
	Friends outside village	1	11	5
	Total	100	100	100

Source: Field survey, 2012.

Notes

* Only the amount of donation made to an identified recipient is given here. The amount of rice is given as paddy (unhusked rice) equivalent.

** In the case of first-generation settlers, the recipients are their children, whereas in the case of the second generation, the recipients are their parents.

*** Other relatives include sisters, cousins, uncles, aunts, and so on.

among the Southeast Asian peasantry is nowadays dominated by "balanced reciprocity," in which each party reciprocates equivalents within a narrow time frame, whereas in the African context a form of "generalized reciprocity," in which a counter-obligation is not always necessary, continues to prevail.

Researchers of Southeast Asian rural communities have pointed out that food sharing or gift giving is generally characterized by people's propensity for balanced reciprocity (Scott 1976: 168). In his detailed description of ritualized food exchanges such as *slamatan* and *jagong* in central Java, Sekimoto (1976) demonstrated that villagers always give careful consideration regarding how to balance the amount of food exchanged between two parties, notwithstanding the existence of economic disparity between them. Focusing on reciprocal labor and gift exchanges (*ao raeng*) in Central Thailand Tsuruta (2004) shows that, along with the penetration of wage labor, the system of cooperative labor exchange underwent a significant change with villagers becoming more calculating and economic, demanding the return for labor on stricter terms. This "calculated reciprocity" was also observed in family celebrations such as weddings and Buddhist ordinations when gifts (mostly cash) were exchanged among the villagers. Except for occasional ceremonial feasts, rice was rarely shared among post-Green Revolution Thai villagers. For example, in one village in northeast Thailand in the 1990s, after some of the Green Revolution technologies had finally arrived, on average only 0.8 percent of the rice produced by each household was donated to other households (mostly living outside the village), while as little as 0.6 percent was consumed in ceremonial feasts (Nakada 1995: 531–4, 1996: 611–16).

Meanwhile, the norm of reciprocity in African villages is still tightly connected with people's subsistence needs. According to Sugiyama (1987), when hybrid maize cultivation was introduced in the Bemba villages of northern Zambia in the 1980s to promote modernized cash crop production, it involved the use of chemical fertilizers in semi-permanent fields in contrast to the existing shifting cultivation of indigenous finger millet. This led to a disparity between villagers who could afford to have maize fields and those who could not, especially female-headed households. However, as she discusses further in Chapter 9 of this volume, when villagers started to replace millet with maize for home consumption, it took on the connotation of being "food" and thus something that could be shared and serve as a means of reducing the growing economic gap among households. The conversion of maize from a cash crop into foods reactivated the sharing practices, which are such an essential component of the inherent "levelling mechanism" built into village communal relations. Behind this phenomenon lies a moral principle that richer households should help poorer ones, especially as regards food, a generalized reciprocity norm, according to which a return gift of equivalent value is not always necessary. Equivalence in such exchanges is "relegated to secondary consideration because what matters most is to get or to offer the necessary amount of something needed" (Tsuruta 2008: 47).

Such is the system of food sharing that was observed in the rice-growing community at Mwea. In MIS, most settlers share a local norm according to which the rich should give food to those who are poorer without seeking a favor in return. Behind such sharing practices may lie a sense of guilt among the first-generation settlers, who feel that they are unable to perform their fatherly duty of preparing enough farmland for their sons due to strict NIB management rules. At the same time, second-generation settlers contribute to the institutionalization of this form of exchange by openly displaying jealousy toward their fathers because they own more land and rice. This seems to awaken a conscience of guilt in the fathers that is only eased by leveling the amount of rice consumption between them, at least in the short term. Through this interaction, the father enjoys authority over his son, while the son and his family can overcome the food shortage. The son may display his gratitude and respect and signal his allegiance, expecting a future inheritance of more *maeka*. Such sharing opportunities embody the ideal father-son relationship in the Kikuyu moral code, which assumes that sons should accord the greatest respect to their fathers, who in turn should help the former in a dignified manner.

What we have observed in the Mwea Scheme is quite different from the cases of Southeast Asia with its more rigid social division between rich and poor villagers. Tenant farmers there are more closely bonded to wealthy landowners in exchange relations that are potentially explosive as they must pay high farm rent and display their allegiance to the landowners at both public and private occasions. For example, already in 1950s Philippines, the landed class was firmly established and 41 percent of surveyed households were tenant farmers who had little choice but to accept their subordinate position. The landowner, for his part, may have extended assistance to the tenants in cases of emergency but it was done as a paternalist gesture rather than as something growing out of a local norm of generalized reciprocity (Anderson 1964: 176–80; Umehara 1992: 158–9, 212–15). In MIS, the social division between land-owning villagers and other smallholders is mollified by the common food-sharing practices that exist because richer settlers are always under social pressure to donate rice to poorer households without expecting anything in return. This practice may at least to some extent hold back wealth accumulation. As noted above, the "selling rate" of rice for first-generation settlers – the relatively larger rice growers – is lower than that of small-scale second-generation settlers, the exact opposite of cases in Southeast Asia, where the gap between the landed and the landless is much more pronounced.

Conclusions

It is possible to draw three main conclusions from the account of livelihood patterns in the MIS. The first is that the trajectory of development is almost in all respects the opposite to what happened in Southeast Asia. Before its Green Revolution rural livelihood patterns were similar in the sense that a "subsistence ethic" underlay rural communities in both places. Modernization of rice farming between the 1960s and the 1980s, however, had the effect of generally undermining existing social security arrangements, which among other things, had allowed tenants and landless villagers to survive through the sharing of rice (or cash) by wealthier villagers. At the same time, reciprocal relations between households became more calculating. The value of what was being exchanged had to be assessed more precisely and within more precise timelines. A "balanced reciprocity" took the place of the more generalized types of exchange that had prevailed before the Green Revolution. Although commercial rice production at Mwea has continued to expand during the same time period, it has not marginalized the moral economy but instead has had the opposite effect of activating mutual aid practices. Liberalization of rice production and distribution exposed settlers to the risks of price fluctuations and water shortages. Their response was to fall back on mutual aid norms and practices which meant that less rice was being sold and more shared and used for self-consumption.

A second conclusion is that even in a highly commercialized peasant community like MIS, tenacious moral values are still alive in everyday social relations among villagers. Traditionally, sharing foods and co-eating among members of a paternal family were important practices in Kikuyu society. Eating alone, refusing to join co-eating events, or accumulating personal wealth was avoided because such behavior denoted the acts of a witch doctor (Kenyatta 1978: 1–19, 98–129). Even if not all settlers adhere to this co-eating philosophy today – some wealthier villagers sell most of the harvested rice and convert it to cash in order to use it for themselves and their respective family – they do so at some cost. They are regarded as unsociable and have a smaller number of ritual kindred than other more generous villagers. The prevailing moral standard continues to be the virtue of mutual aid as manifest in the ideal father-son relationship, in which the former must prepare enough farmland for the latter to use. The uprising by the second-generation settlers in 1998 seems to have enhanced the feelings of guilt of the first-generation settlers, prompting them to distribute rice to their sons in times of hardship. The same moral standard is present when second-generation settlers in the *jua kali* area collaborate with each other to bring out the best possible harvest from self-made but labor-intensive paddy fields.

Recently, the Japanese International Cooperation Agency (JICA) has been promoting further commercialization of agriculture at Mwea and in other areas of Kenya by making the marketing system more effective through such measures as adjusting the timing of shipments according to price fluctuation (JICA 2015, 2018). It is questionable, however, whether the mere promotion of "smarter" sales of rice and other cash crops will lead to an improvement of social well-being for the settlers. Such a development project does not consider the fact that settlers may have their own standards of welfare and reasons for continuing to do what already works at a local level. As we have seen in this chapter, from the settler point of view, surplus rice must be shared among neighbors and relatives. Such an act is highly appreciated and gives both social status and power. In contrast, selling one's rice is the fate of a miser or someone who cannot cater for himself. If they store all the paddy in the post-harvest period waiting for the price to rise in the off-season, they may miss the opportunity to have a *mucemanio* event, the most important occasion for cementing social relationships among neighbors. A mere emphasis on maximizing profit also threatens the existing informal system of subsistence security, which is still important to settlers in MIS because a formal crop insurance system has yet to be introduced. The final lesson, therefore, is that policymakers should pay closer attention to how local African peasants live and not destroy their unique system of redistributing wealth.

Notes

1 This section is based on Ito's fieldwork carried out between 2012 and 2017. A portion of the research results have already been published in Japanese (Ito 2016, 2017a, 2017b).

2 In terms of legal ownership, the settlers were given only usufructuary rights, not proprietary rights of the farmland (Ishii 2007: 128).
3 The basmati variety in Kenya is an improved variety of the basmati rice originally introduced from Pakistan and it is quite responsive to fertilizer application, hence high yielding (Njeru et al. 2016: 312).
4 One bag contains between 90 and 100 kg of unhusked rice.
5 As all sample farmers are Catholic, the ritual kindred system is observed on ceremonial occasions such as baptism, confirmation and marriage. At baptism, for example, one of the respected elders may be asked to become a godfather (*baba ya ubatizo* in Kiswahili) of the baptized children. The godfathers are required not only to give religious advice to their godchild but also to offer some form of social security to the child and his or her family.

8 Longstanding imprints of natural society

Toward a new science of agriculture in Africa

Kazuhiko Sugimura

Introduction

The purpose of this chapter is to initially pull together the main findings of previous chapters in this volume and add some new materials drawn from research among the Kumu people in the Congolese rain forest. This first section will demonstrate the persistence of Natural Society features that make much of the agricultural scene in Africa so different from what is found in other regions of the world. The second part of the chapter is devoted to comparing rural livelihoods in Africa with those of other regions that have historically gone through the transformation to an Agrarian Society. The analysis centers on the economy of sharing and the notion that life is organized around consumption and reproduction rather than production. This type of livelihood system can be found across the sub-Saharan region of Africa. As the content of this volume illustrates, it encompasses dry savanna areas inhabited by agro-pastoralists, and tropical rainforests along with their woodland fringes dominated by swidden cultivators. These are the areas that we refer to as "the South of the South," in the sense that they are the least developed areas in Africa. They also constitute the "forgotten" part of Africa which continues to drag itself along without the dynamic that commercial agriculture fed by Green Revolution technologies has created in areas with favorable climatic and socioeconomic conditions. Thus, rural Africa is not a homogenous entity. Regional disparity has grown in recent years and extensive areas of sub-Saharan Africa remain untouched by policy efforts to modernize agriculture. Experts and policy practitioners alike have promoted measures that make sense in countries which have already undergone a transformation to Agrarian Society but fail to address what it means to develop agriculture in a Natural Society setting. That is why this chapter, and the volume at large, is pointing to and arguing for a new science of agriculture in Africa.

This does not mean that these forgotten parts of Africa have escaped the cash economy. Since the 1990s, even the most remote areas in Africa have been extensively affected by the wave of globalization, especially the penetration of a cash nexus into village social relations. The preceding four chapters have shown how the lives of hunter-gatherers, shifting cultivators, and agro-pastoralists, not to mention commercialized rice growers, have become immersed in a cash economy. What they also show, however, is that this immersion does not necessarily imply the disappearance of local idioms of organizing social life. Instead of a break with the past, the presence of cash is fused with existing consumption communities. Surplus may be used in a rational manner, but the reasoning is different. It is not applied to promote greater productivity but rather to secure subsistence for all by diversifying income and sharing food. This consumption perspective has survived the major socioeconomic transformations attempted in the colonial and post-colonial eras. By still being a prominent feature of rural Africa, it stands out as quite different in a comparison with other regions of the world.

The economy of sharing

As we noted in Chapter 2, research questions from the early Kyoto Africanists revolved around the interactive relationship between humans and nature. As they tried to pursue the origins of human society, they focused on hunter-gatherers, pastoralists, and shifting cultivators. They revealed in their detailed fieldwork that these peoples do not try to modify nature for their own sake; rather, they adapt themselves to the given environment. Rural dwellers in Africa devised a variety of ways to cope with the vagaries of nature, especially by adopting livelihood methods with mobile characteristics (Tanaka 1982). The high mobility of the inhabitants is also associated with their flexibility in choosing a means of subsistence such as hunting, gathering, pastoralism, and shifting cultivation. From a sociological viewpoint, it is notable that they developed fluid types of communities that could be flexibly reorganized according to changing circumstances and surviving an unfavorable environment by utilizing the network of interpersonal relationships. Their livelihoods are organized in an "economy of sharing" which binds close kin together through the medium of everyday exchanges of foods and other valuable items. At times of scarcity this principle of sharing is voluntarily extended to include others in need: no one should go hungry!

Communal eating practices among the Kumu people

Under the influence of the Kyoto field research tradition, the author carried out an anthropological research project among the Kumu swidden cultivators in the rainforest of what in the 1980s was Zaire and now is referred to as the Democratic Republic of Congo (DRC). I found intriguing communal eating practices among the Kumu and neighboring ethnic groups, which provided an important opportunity to cement social bonds among the kin (Sugimura 2004).

Each communal eating group of the Kumu is made up of five to ten households, members of which belong to the same paternal lineage group, *toa*. Co-eating is held twice a day, morning and evening, in which women of each household prepare and bring the meals (a starchy staple with an accompanying dish) to share. Men, women, and children are separated and eat together within each group, enjoying a variety of meals. The practice of co-eating not only equalizes the food consumption of each household but also eases the women's burden of preparing meals every day. In addition, the practice has a variety of social functions. Most notably, this is an important occasion for communication between *toa* members, at which decisions about communal matters are made. In this system, access to food is guaranteed to the underprivileged villagers, including those who cannot work due to disease or accident. The co-eating group also takes care of orphaned children who are too young to live alone.

The communal eating practice of the Kumu contributes to forging a sense of community and to reproducing social relations based on the economy of sharing and mutual assistance. *Toa* members tend to help each other on various occasions, notably in farming activities, caring for sick persons, settling disputes, and preparing for ceremonies including weddings and funerals. Wealthier members of the same *toa* tend to spend more money on and provide more food to the co-eating event. Behind such behavior lies the fear of witchcraft, a parallel to what was found among the Tongwe discussed in Chapter 5.

The *toa* is not a fixed, stable group but relies for its persistence on the ability of members to participate actively in the everyday co-eating practices. Without such an occasion and with fewer opportunities to reconfirm their membership of the same paternal lineage, the group could have become more unstable and even broken up. *Toa* members uphold the tradition of co-eating as a means of confirming their common identity. The notion of being together does not rest on the tangible basis of productive resources like scarce land and water as in Agrarian Society. Instead, it persists as a social construct with origin in the notion of a community of consumption.

A compelling question is how it has been possible for the economy of sharing in rural Africa to survive the forces of capitalism and globalization, which grew stronger especially after the turn of the century. A look back at what has been revealed in previous chapters in this volume helps answering that question.

The persistence of sharing

One of the most interesting aspects of social change under way in contemporary rural Africa is that, despite an apparent similarity with other developing countries in the South, there is still something quite distinct about the local communities in Africa. As demonstrated in Chapter 7, even in the technologically advanced and highly commercialized farming communities of Kenya, the ethic of sharing foods remains a common practice for people to stay connected and survive acute food shortages. The culture of sharing among the San hunter-gatherers, as discussed in Chapter 4, has persisted even after the resettlement policies were implemented (Osaki 1990; Maruyama 2003). Interestingly, cash, which is normally considered as something not to be shared, is being used among agro-pastoralist villagers, as discussed in Chapter 6, to exchange for sharable goods or for opportunities to obtain wage labor.

The presence of wage labor is normally considered as evidence of the penetration of capitalist relationships and social differentiation, but such a notion is too simplistic and misleading as wage labor in African villages may have the function of sharing scarce cash among households. In Tanzanian villages, for example, the author observed that wage labor (*kibarua* in Kiswahili) is sometimes arranged by wealthier villagers in response to the request of those villagers in desperate need of cash (Sugimura 2006; Maghimbi et al. 2016: 220–1). The wealthier person will not refuse the request as he wishes to maintain his respected status in the community. *Kibarua* provides a means of sharing cash that is indispensable for villagers to survive. The relationship between hirer and hired is not a fixed one and tends to change from case to case.

The same sharing ethic has also been observed in rural finance. When commercial production of maize was introduced to northern Zambia in the mid-1980s, villagers were able to obtain necessary inputs (hybrid seeds and chemical fertilizers) on credit through the agents appointed by the government. In a Bemba village where the repayment rate of the loan was very low, people were warned that those who failed to repay 75 percent of their debt would be excluded from the lending scheme. Faced with this threat, some wealthier villagers with cash income managed to pay back their loan in full, which eventually pushed up the repayment rate of the whole village and convinced government agents to extend credit to all debtors for the next year. Interestingly, the wealthier villagers never criticized their poorer neighbors for failing to meet the credit. They viewed it as entirely right to refuse the repayment, because it was a given that poorer households should first use the harvested maize for domestic consumption. This episode shows that, although villagers regularly act out of self-interest, they still share the basic principles that (1) securing subsistence should be given a high priority and (2) the haves should pay more than the have-nots (Sugiyama 2001).

Communities of consumption and production

When analyzing village communities in Agrarian Societies, the focus is often narrowed down to the villager's right of access to the means (or object) of production, e.g., the distribution of land or water resources necessary for farming activities. Thus, more attention is paid to the "labor-production process" than to the "consumption-reproduction process," although the latter might be as important in building interpersonal relations and mutual trust. The result is that African village economies have been studied from a narrow economistic view.[1]

As we have shown in this volume, the organizing principles of communities in rural Africa are not necessarily associated with collective labor processes related to agricultural production. For example, in Kumu villages, once the land for cultivation is allocated among members of the same *toa*, swidden fields are basically managed independently by each household. Among these people and neighboring ethnic groups, there are cooperative labor exchange practices in clearing short-fallow bushes, which require relatively light labor. However, the hardest and most dangerous work of clearing primary forest areas is usually done individually (Sugimura 2004). Cooperation between households is embodied in the "consumption process" rather than in the production process. Villagers interact by visiting each other, eating together, or exchanging gifts, in order to build up the mutual trust that is indispensable for stable social reproduction. Above all, the co-eating practice offers the most important occasion for strengthening family cohesion. Social relationships between individual villagers are based on how long and how often they can eat in common with other community members.

Livelihoods in rural Africa, therefore, are pursued in the context of communities of consumption where the social logic is based on the principle of sharing and especially the notion that the rich have a responsibility to secure the subsistence of their poorer brethren. In these communities, human reproduction matters more than material production. The consumption communities in rural Africa tend to be flexible units that often split or unite in response to population changes making this type of social unit difficult to capture by the state or by market-driven incentives for profit (Hyden 1980). They follow a logic that does not deny individual households from engaging in various commercial activities – like diamond and gold mining among the Kumu – but each individual household is not just an independent economic unit, which seeks commercial gain for its own sake, but rather part of a wider community of sharing among close kin. Because land is a communally controlled asset no social differentiation arises from unequal access to land. Although the increase in off-farm activities tends to create differences in wealth, all villagers are roughly equal in the sense that as subsistence peasants they continue to share responsibility for each other. The African experience invites a comparison with Chayanov's analysis of the Russian peasantry a hundred years ago.

Beyond Chayanov's model

Alexander Chayanov, an agricultural economist in post-revolutionary Russia, is the first scholar to theorize about the organizing principle of the peasant unit of production as distinctly different from the capitalist farm enterprise (in English translation Chayanov 1966). One of his key findings is that the production level (in terms of farm size and utilization of domestic labor) of a peasant household tends to be determined by family size or the consumptive demand of all household members. When equilibrium between production and consumption is achieved, a portion of family labor may be left unused ("underuse of labor power"). As an indicator of the changing labor-consumer balance, he devised a coefficient C/W (consumer-worker ratio), which changes according to the growth cycle of the peasant family. Chayanov made an important contribution by viewing the peasant household not only from a production but also a consumption perspective. His model, which was formulated upon the basis of his empirical research in Russia, is relevant for contemporary Africa but it cannot be applied without modification. In Chayanov's model each household is treated as an independent social unit while in Africa, as we have shown in this volume, households come together in communities of consumption for the purpose of securing subsistence and reach Chayanov's demographic equilibrium. Labor intensity, therefore, is assessed in a communal setting as, for example, in the co-eating group among the Kumu. The farm size and the location of clearing swidden fields are basically determined by each such group. The C/W ratio within each *toa* has a considerable influence on how individual households utilize labor. Thus, *toa* units with a high C/W ratio tend to clear a long-fallow primary forest in order to get more foodstuffs and feed more mouths, while those with a low C/W ratio tend to clear a short-fallow forest, which requires less labor input. Within the *toa*, however, the farmland size of each household tends to be equalized (Sugimura 2004). Balance between production and consumption is achieved not within individual households, but among all *toa* members who are related to each other through interfamilial sharing of foods.

Accumulation in African villages

The importance of reproduction in African villages provides a reason for considering the meaning of accumulation in rural Africa.[2] A valuable starting point for doing so is the work of the Japanese economist, Hisashi Nakamura (1976). According to him, accumulation is a process in which a community attains a higher stage of production on the foundation of the fruits of past labor – the collective achievements of its ancestors. Accumulation of labor is made possible by the improvement of any of the three incommensurable elements of the labor process: (1) subject of labor (human labor); (2) object of labor (arable land); and (3) means of labor (tools and instruments). Based on this division, he further formulated a typology of accumulation of productive resources (Figure 8.1). The arable land type of accumulation is built up through efforts to raise land productivity. The instrument type of accumulation is a process defined by the efficiency

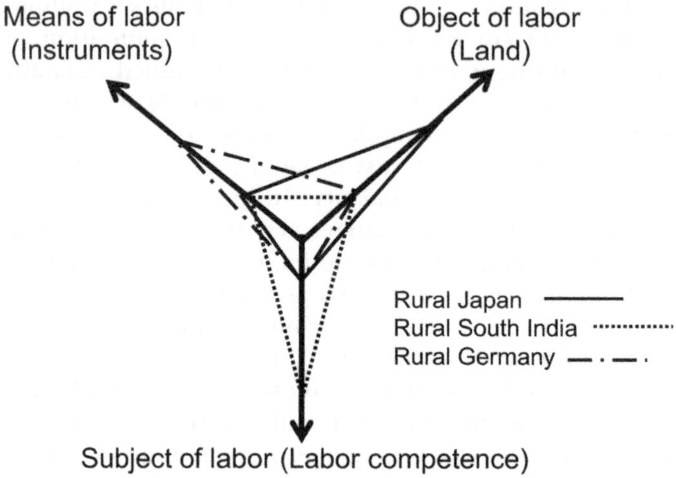

Figure 8.1 Three types of accumulation.
Source: Nakamura (1976).

of farming tools and machines. In the labor competence type of accumulation, the result of labor is determined by knowledge, experience, information, labor skill of individual peasants, as well as an efficient labor arrangement such as cooperative work or division of labor. Here, labor arrangement is regarded as a social extension of labor ability beyond the individual human body.

Each Agrarian Society combines these three accumulation types in different ways leading to significant variations in social structure. According to Nakamura, agricultural development in Europe (Germany) is characterized by the improvement of tools and machines (the instrument type of accumulation). In Japan, by contrast, increase in agricultural productivity has been achieved primarily through long-term investment in land (the arable land type of accumulation). Adding southern India, where Nakamura carried out fieldwork, the labor competence type of accumulation is predominant through the division of labor based on the caste system.

The subsistence economy in rural Africa, in which social relations are far more valuable productive resources than land and tools, is best characterized as an example of the labor competence type of accumulation. Yet, it has features of its own because accumulated wealth is invested in reproductive activities rather than in production. Referring to precolonial Tanzania, Koponen notes about peasant life:

> It was not that accumulation was missing but rather that what was accumulated was not used to lay a material basis for future economic

> and social activity. Instead it was used to create new social relations and new human beings and, in this way to lay a social basis for that activity. This, I suggest, can be regarded as the 'innermost secret' of these societies.
>
> (Koponen 1988: 389)

His view is quite similar to the idea of "wealth in people" (as opposed to "wealth in things") which has been applied in historical and anthropological studies in equatorial Africa (e.g., Miers and Kopytoff 1977; Guyer 1993). The concept of wealth in people is suggestive in understanding the mode of accumulation in African rural communities, in which wealth is accumulated primarily in the form of livestock. Livestock is not only important as a means of subsistence but is also used as a form of currency and bride-wealth. For villagers, marriage is one of the best opportunities to extend their social networks and thus to expand the scale of their economies.

While livestock is generally considered as a means of agricultural production in Agrarian Societies, it plays a very important role in creating social bonds through gift giving (typically in relation to bride-wealth) among pastoralist and agro-pastoralist communities in Africa as discussed in Chapter 6. Even among shifting cultivators in rainforests without cattle like the Kumu and neighboring ethnic groups, goats are considered a symbol of wealth and thus used as bride-wealth. The number of goats in one's possession is the most important criterion for judging whether a person is wealthy (Sugimura 2004). Agro-pastoralists, pastoralists, and even shifting cultivators in Africa tend to use accumulated wealth to purchase livestock or to marry a second wife rather than to invest in improving arable land or farming tools. African village communities are unique in that human-human relations take precedence over human-land relations. Villagers have less attachment to the land, compared to peasants in Agrarian Societies who are tightly bound to the land, and differentiation in ownership of land has long been the major source of wealth and stratification of society.

Comparing Africa to Southeast Asia

To elucidate the uniqueness of the African peasantry, it may be helpful to once more compare rural societies in Africa with those in Southeast Asia. Using the accumulation typology introduced above, Tomosugi (1980) asserted that in traditional Thai peasant communities, human labor, not tools or land, was the most important element. For example, in central Thailand where land was abundant and there was no significant development of farming tools, the primary preoccupation for peasants was to secure the necessary labor force through dyadic social networks among relatives and friends. He associated these networks with the notion of "loosely-structured social systems" (Tomosugi 1975), a well-known thesis describing Thai society as lacking the cohesion and cooperation among households that had promoted development in regimented societies such as Japan (Evers 1969). With the benefit of hindsight, we now know that commercial rice production for export in Thailand gradually proliferated to involve subsistence peasants. By the 1970s, Green Revolution technology had penetrated central Thai villages. Through this whole process, Tomosugi (1980: 80) asserts that the accumulation type of labor shifted from being initially about labor competence to being about arable land, and finally about technology when "tractorization" took off in the 1970s. He also argues that, with this transformation of accumulation practices, the mode of ownership, particularly of land, was changed from "occupancy" to "private ownership," and former reciprocal exchanges between villagers were similarly replaced by profit-minded economic exchanges (ibid.: 82–7; Tsuruta 2004).

The point here is that preindustrial Thai rural communities were quite similar to their counterparts in Africa in that both were loosely structured societies, characterized by a lack of collective labor organization tied to a specific piece of land. But why did the rural communities in Thailand transform themselves into commercial rice production units by adopting new technologies, while those in Africa have remained subsistence-oriented despite government efforts to modernize agriculture? The answer is that behind these divergent paths lies the fact that rural communities in Southeast Asia had largely been communities of production rather than consumption. According to Tsuruta (2004: 107), households in rural Thailand, mostly made up of just a nuclear family, may have long been established as independent units of production and consumption. The household, as a self-reliant economic unit, has considerable leeway in undertaking commercial activities with little communal bondage to prevent it from engaging in economically rational action. This is in contrast with rural Africa where households as social units are more fluid, and individuals are connected to each other through the institutionalized exchange of consumptive goods that serve to level economic inequality.

Such leveling mechanisms did exist in rural Southeast Asia, as can be seen in the phenomenon which Geertz (1963) referred to as "shared poverty." In rural Java, villagers shared available economic resources and opportunities by further fractionalizing an already dwindling economic pie. This practice, however, differs from what exists in Africa today because the leveling was carried out as a function of production, not consumption. The result was that by the 1970s, Java villages had a considerable number of landless members, who depended entirely on wage labor, in contrast to African villages where economic disparity stemming from unequal land ownership has traditionally been non-existent. The nature

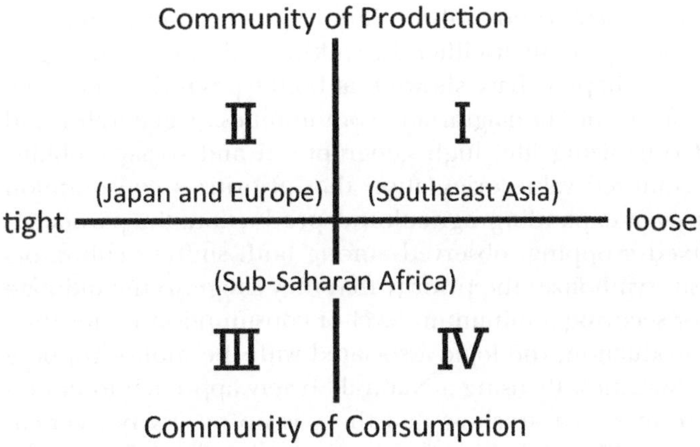

Figure 8.2 Four types of rural communities.

Source: Author.

of Southeast Asian villages as communities of "agrarian" production can also be observed in the value of livestock. For example, farmers have traditionally kept water buffaloes for ploughing their paddy fields but by the 1980s they had virtually disappeared in central Thai villages and been replaced by tractors. In rural Africa, on the other hand, livestock has continued to serve not just as a means of production but foremost as a source of social wealth.

Africa and Southeast Asia may be similar in that both are loosely knit societies, as compared to tightly knit societies like Japan and those in Europe. Nonetheless, as we have shown, they differ in their social formations – Southeast Asian societies being communities of production while those in Africa are communities of consumption. Using these distinctions, rural communities can be organized into four types as illustrated in Figure 8.2: (I) loosely-structured societies based on communities of production; (II) tightly-structured societies based on communities of production; (III) tightly-structured societies based on communities of consumption; and (IV) loosely-structured societies based on communities of consumption. The point we wish to make here is that from a comparative perspective, most villages in sub-Saharan Africa remain communities of consumption (quadrants III and IV) which may hold back progressive individuals from responding to market incentives but which at the same time serve as leveling mechanisms in society. Social reproduction does not only involve reproduction of the labor force, as Meillassoux (1975) argued, but also the community of consumption. These are the informal institutional structures that shape social formations in African society.

Toward a new science of agriculture in Africa

In this volume we have focused not only on hunting and gathering but also two types of traditional peasant livelihoods: shifting cultivation and agro-pastoralism. Preceding chapters have shown that both types reflect the value system and mode of life of "non-agrarian" communities: a generalist and "extensive" way of organizing life, high geographical and social mobility, and a subsistence-centered value orientation that prioritizes consumption and reproduction over expanding agricultural production. In particular, the practice of mixed cropping, observed among both shifting cultivators and agro-pastoralists, symbolizes the priority that they assign to the ultimate peasant principle of securing a minimum level of consumption rather than the maximum of production, the logic associated with the mono-cropping system of Agrarian Societies. By using a Natural Society approach to understanding agriculture in Africa we show that there is no necessary clear-cut distinction between "cultivators" and "herders" nor is mixed farming in Africa pursued in the same way as, for example, in Europe. Communities of consumption in Africa generate a very different logic than those that are created around production – a distinction that has yet to be fully recognized in mainstream research and policy circles. The search for a science of agriculture that suits Africa or "the South of the South," as we also call it, has just begun. Three questions seem especially important for guiding that exercise: (1) why should we bother now? (2) what evidence exists for a new science? and (3) how might it be carried out?

The timeliness of a new science

Efforts to change peasant agriculture in Africa are over a century old. It began in colonial days and has continued postindependence by governments advised by donor-funded experts. They have typically approached their task from the perspective of a society that has already undergone transformation from Natural to Agrarian Society. For example, colonial and postindependence administrators have favored settled agriculture over nomadic pastoralism, ignoring the risk-spreading merit of pastoral or semi-pastoral livelihoods in arid environments. They have had a constant preoccupation with the introduction of the English model of mixed farming, although most attempts to facilitate integration of livestock husbandry and farming in Africa have proven to be a failure (Sumberg 1998). This model is in fundamental conflict with the land-use pattern of indigenous cultivators, characterized by great mobility and the absence of private ownership of land. Instead of trying to build on already existing knowledge among Africa's peasant cultivators, the mainstream expert advice has been to replace it with practices that may be proven in Europe and North America but are untested in the African context. Enough time has now passed that in the light of the many shortcomings associated with this approach there is good reason for trying to develop African agriculture from the bottom up rather than from the outside in.

The issue has taken on new urgency as global food security is being threatened by climate change. The challenge here is to think globally but act locally. The grand vision of global food security needs to be aligned with the pedestrian yet fundamental peasant notion of subsistence and food sharing. While the urgency of the international community is justified, it may backfire without enough understanding of local conditions and peasant rationales. There is already impatience in many policy circles and a readiness to abandon the small-scale peasant sector in favor of large-scale commercial agriculture. Alienating communal land for private ownership by corporations or individual investors is now causing social conflict in many African countries, a trend that may get worse as land becomes an increasingly scarce resource. The issue of peasant agriculture, therefore, needs to be viewed in its broader social and political context.

The evidence for a new science

Mixed cropping is closely associated with peasant values, which attach great importance to diversity and stability. For example, like many other peasant cultivators, the Kumu villagers view mono-cropping in negative terms associating it with both risk and greater labor-intensity. They prefer mixed cropping as it provides a variety of crops of various nutritional values throughout the year, leading to a stability of their livelihood. In other words, they seek a "diverse" rather than an "abundant" harvest. Long fallow is not treated as a problem but rather as an opportunity for growing a greater variety of crop species and thereby achieving diversity and sustainability of the farmland ecosystem (Sugimura 2004).

The point is that mixed cropping works. Although the yields in the fallow-mixed cropping system as practiced by shifting cultivators vary tremendously due to local climatic, topographical, and soil conditions, it is not necessarily an unproductive method. For example, Ankei (1981) calculated yield per unit area of three crops (cassava, plantain, and rice) mixed in swidden fields among the Songola in the rainforest of what was then Zaire. According to his estimate, the minimum yield per hectare of cassava was 9,000 kg, that of plantain 3,000 kg, and that of rice 800 kg.[3] In terms of calories, one hectare of swidden fields thus produced 12.6 giga-calories in total, exceeding the 11.3 giga-calories produced on one hectare of monoculture paddy fields in Japan (Sugimura 2004: 255).

Richards (1985) highlights the amazing diversity of agricultural practices, which stem from the fact that peasant cultivators accommodate their technique to the variable environment instead of creating a uniform environment favorable for farming such as making terraced fields or irrigation facilities. This adaptability to existing environments reflects each cultivator's local knowledge and understanding of the conditions for farming in specific areas. This approach to farming leaves a light footprint on the land but does not preclude innovation. On the contrary, there exists a variety of preindustrial methods of intensification such as fallowing, crop rotation, use of nitrogen-fixing crops, manuring, composting, and small-scale irrigation (Djurfeldt 2005: 10–11). These innovations, however, are largely ad hoc and piecemeal and have never coalesced into a standardized farming system. The farming system of each specific locality is the result of a well-crafted adaptation to the existing natural environment through conscious selection of crops and seeds as well as the utilization of topographical conditions including natural waterways. As one historian points out, in contrast to standardized Green Revolution technology, "African peasant farming was a skilled craft producing numerous crops adapted to small variations of soil and climate" (Iliffe 1995: 268). Our volume proves that this craft has not been lost.

Others have provided similar evidence. Although increased population pressure and penetration of the market economy have prompted villagers more recently to adopt intensive farming techniques, polyculture and diversification to secure subsistence still characterize agriculture in many parts of Africa. Connelly and Chaiken (2000) reveal that, in contrast to other regions, intensification of agriculture in Africa does not always lead to crop specialization and a loss of agro-diversity. For example, in western Kenya, as villagers intensify their farming systems, they enhance diversity through such measures as intercropping, poly-variety and integration of crops and livestock. Another study on agricultural intensification in a densely populated area of Kenya confirms that intensification does not necessarily result in a landscape of monoculture (Tiffen et al. 1994).

There is plenty of evidence that the indigenous forms of agriculture in Africa are productive without being driven by scientific findings. This is an important acknowledgment at a point when it is evident that the uniform application of modern technologies, e.g., the introduction of chemical fertilizer and farm machinery, has largely failed to bring expected results. Rather than opposing or destroying indigenous farming systems and practices there is reason to develop them. The case of the Nupe in Nigeria is instructive. They have long cultivated rice in the lowlands during the rainy season on semi-paddy fields, in which ridges are pulled down every time after harvest to plant dry-season crops. In the 1990s, a team of Japanese researchers attempted to introduce Asian-type large-plot paddy fields with permanent ridges to increase land productivity. Villagers, however, were very reluctant to accept them, apparently because permanent ridges would have prevented them from practicing mound cultivation in the dry season. They stuck to existing farming techniques which combine cultivation both upland and lowland, where they apply several different methods of land preparation, each being well adapted to the respective microenvironment and topographical condition. The notion of permanent paddy fields also ran into conflict with the existing land tenure system, in which the right to use land was flexibly exercised (Hirose and Wakatsuki 1997). Their aversion to be bound to a specific piece of land may also have stemmed from the fear of losing flexibility in their way of life. Given the precarious nature of dryland agriculture, to be tied to a given piece of land is tantamount to losing the flexibility of dealing with unforeseeable environmental changes.

Ways of moving ahead

Institutionalized agrarian sciences are not free from the ideology of Agrarian Societies. Most formulas of rural development are based on the experiences of ex-Agrarian Societies and cannot be applied directly to sub-Saharan Africa, which is marked by a series of "non-agrarian" features. One such formula is agriculture-centeredness, particularly the sole focus on agricultural intensification and monoculture. Although African rural communities are rarely specialized in agriculture, as seen in agro-pastoralist societies, rural policies always attach disproportionate importance to modernizing agriculture (Maghimbi et al. 2016).

Africa needs a new agricultural science which is free from this "agrarian bias" and leaves enough scope for initiatives from below. Richards (1985) once claimed that, in order to avoid the repeated failures to introduce modern technology from above, agricultural scientists and extension workers must pay special heed to "people's science" at the grassroots. Remarkable indigenous techniques such as shifting cultivation and mixed cropping demonstrate the great ingenuity of African peasants to adapt to the given environment, which may change from one year to the next. These locally developed technologies fit well into diversified rural livelihoods and heterogeneous natural environments. They contribute to environmental conservation and the long-term economic security of local communities. These measures may not amount to an agricultural revolution but to a way of grounding rural development in local practices. They constitute an alternative to the Green Revolution characterized by its top-down approach and one-size-fits-all application of the scientific method. Thus, the development of African agriculture should not be viewed solely in terms of technical changes but rather be examined as a part of wider social process which entails changes in values and the mode of consumption, an approach that will be laid out and further explained by Yuko Sugiyama in Chapter 9.

Conclusions

The international community is increasingly concerned about taking measures that will reverse the deterioration of nature and find ways of inventing systems that secure a sustainable use of the earth's dwindling resources. The South of the South, i.e., sub-Saharan Africa, should be the obvious central point for this search. The footprint of indigenous agricultural practices is light and adjusted to local conditions that promise a sustainable development. "Subsistence" is becoming an integral part of a new paradigm for human development, which recognizes the value of the long-term security of communities rather than the short-term economic growth of each nation (Bennholdt-Thomsen and Mies 2000). Rural Africa is home to the world's largest and richest subsistence cultures. The approach to rural development in Africa must be based on the "non-agrarian" features of its people and their livelihood patterns, characterized by "extensive" modes of subsistence and a unique mode of consumption and reproduction. Viewing agricultural development in its broader social context invites consideration of the many non-agrarian aspects of life that shape land use in Africa. Because Africans are organized in communities of consumption, they have a sense of progress and production not as ends in-and-of-themselves but as means to secure subsistence and food security. This philosophy is consistent with prevailing notions of environmental conservation but largely ignored because it lacks the coherence and profile of international panel statements on key global issues. Researchers need to dig deeper to ensure that the many secrets of how rural livelihoods in Africa are constituted and sustained do not get lost.

Notes

1 Most studies of African agriculture and rural societies in Japan since the 1960s have been made from the viewpoint of agricultural economics, focusing, like scholarship in western countries, mostly on production and marketing of farm products. The ecological anthropologists in Kyoto, in contrast, have tried to elucidate the totality and complexity of the subsistence economy in African villages, including its pattern of consumption that forms the core of village social organization.

2 French Marxist anthropologists, notably Meillassoux (1975), asserted that the system of reproduction of life, notably the kinship system such as lineage, is as important as the system of material production in African rural societies. They were, however, not concerned about how the lineage functions as an everyday livelihood group, and were not free from a dependency theory bias, according to which the African peasantry was viewed as being ultimately subjected to exploitation by the world capitalist system.

3 This is, however, only possible when an ample duration of a fallow period is secured. At the same time, if we make a comparison between the land productivity of swidden-fallow fields and that of permanent fields, one has to take into consideration the length of the fallow period in order to get an accurate measure of land productivity.

9 Making innovation socially inclusive

Lessons from the Bemba in Zambia

Yuko Sugiyama

Introduction

This volume has shown that peasant agriculture in Africa is dynamic but occurs in the context of enhancing social reproduction and consumption rather than production. We have made the point that efforts to modernize agriculture have fallen far short of expectations because they have been based on theoretical premises derived from countries that have gone through a transformation to Agrarian Society. Crop specialization, monocropping and integrated mixed farming systems practiced in countries of the North have been largely rejected by African villagers who see the risks of adopting such practices in their local environments, be they woodlands, rainforests or lands bordering these two types of ecology. This, however, does not mean that African peasants are just risk averse. They are indeed very interested in change but want it, like other people around the world, on terms that they can control.

This chapter explores the uniqueness of the African peasant innovation process, based on the history of innovation among the Bemba of northern Zambia. In the course of our 17-year field research (1983–2000) among these people, we found that the local farming system had been fine-tuned over a hundred-year period of trial and error leading to landmark innovation.[1] Our review of the local folk innovation history shows that new crops and cultivation techniques were constantly introduced and combined with existing local knowledge and techniques in utilizing available resources. As we will demonstrate here, this process has been closely linked to the villagers' readiness to adapt to socioeconomic changes such as the introduction of a cash economy in the colonial era, changing government policies and the formation of a national economy after independence. Changes in farming technology, therefore, need to be examined from the viewpoint not only of on-farm activities per se, but also of off-farm activities and everyday social relations among villagers. It entails studying two sets of mutually overlapping interactive processes between the community and cash economy, and the on-farm and off-farm activities including the sociocultural arrangements of resource utilization.

Bemba livelihood activities

The Bemba are Bantu shifting cultivators living in the *miombo*[2] woodland of northern Zambia. They once formed a powerful chiefdom based on a matrilineal society, whose political system was built into the colonial government system and has been largely retained even after independence. Although the political structure of the Bemba chiefdom is highly centralized, the residential pattern is characterized by dispersion; people are scattered into small villages generally made up of ten to 70 households, which include the village headman and his matrilineal siblings as its core, with a strong uxorilocal pattern of residence. Divorce is frequent among villagers and 30 to 40 percent of village households are headed by females. Because they are shifting cultivators, the village site itself tends to be moved every ten to 30 years.

The Bemba farming system is characterized by a unique method of shifting cultivation, locally called *chitemene.* It is a system well-adapted to the *miombo* woodland environment where conditions for farming are unfavorable because soil fertility is poor and vegetation sparse. Annual precipitation is approximately 1,000 to 1,200 mm, most of which falls in the rainy season from mid-October to March. Cattle-keeping in the villages is scarce because of the prevalence of sleeping sickness conveyed by tsetse flies. From an agronomical viewpoint, the *chitemene* system is characterized by (1) its unique method of preparing the field, and (2) its cropping pattern. When preparing the field, men climb trees and cut off only the branches. After the fallen branches have dried, women carry them to the center of the area and pile them on to a circle-shaped bed. Burning the branch bed makes a *chitemene* cultivation field (in-field), called *ubukula.* This allows trees standing outside the *ubukula* (out-field) to start sprouting new leaves at the cuts left in the tree trunks. Thus, this method has the advantage of not only replenishing enough organic matter into the soil to grow new crops but also promoting regeneration of the woodland.

The cropping pattern is characterized by a mixed cropping and crop rotation system. The crop rotation has a five-year cycle. In the first year, immediately after burning the piled branch bed, seeds of vegetables (typically tomato, local spinach and edible gourd) are sown. Pumpkin and edible gourd seeds, sweet potatoes and beans are planted in the long mounds established on the fringes of the field, after which, cassava stems are planted in the center of the field. In December, when they have enough rain, villagers plant finger millet, their staple food, which is harvested from June to August the following year. In the second year, groundnuts and Bambara nuts are planted and harvested. In the third and fourth years, cassava roots, which have grown big enough to be consumed, are harvested according to villagers' needs. In the fifth year, several round mounds are made to plant beans in an area of the field. After harvesting them, the field is left fallow in order to recover. The regeneration of the forest takes approximately 33 years in the in-field, and 15 years in the out-field environment (Oyama 2013). The crop rotation system allows for intensive use of the fields and the cultivation of a variety of food crops while also protecting the soil surface from solar radiation damage.

The multifaceted pattern of resource use is making the *chitemene* system environmentally sustainable.[3] To borrow the term coined by Kakeya (1976), they are not specialists in agriculture but "generalists" in utilizing the surrounding natural environment in multiple ways. The Bemba villagers utilize the diverse vegetation of the different stages of succession in the *miombo* woodland under fallow, through hunting, fishing and gathering as well as farming. Gathering wild edible plants, collecting edible insects, and hunting and fishing activities provide them with the necessary vegetable nutrients and animal protein.

Their subsistence activities rest on a clear gender division of labor. Men cut tree branches and engage in cash generating activities such as peddling, while women carry the branches to make *chitemene* fields and continuously engage in agriculture thereby taking the responsibility of food management. Women may get money through the brewing and selling of beer, but it is often used to secure male labor for cutting the *chitemene* branches. Difference in gender composition among households is a main reason for economic disparities. Households with greater numbers of males tend to have bigger farmlands for both subsistence and cash generating activities. A lack of male members generally means that they cannot clear the farmland widely enough for subsistence (Oyama and Takamura 2001).[4] Such inequality in the production process, however, tends to be leveled through the consumption process in which foods are shared beyond the individual household unit. The principle of sharing abundant goods (particularly foods) with the have-nots is deeply rooted in Bemba villages. This leveling mechanism also discourages people from expanding *chitemene* fields, because whatever surplus products they harvest are likely to be consumed by other households. This, in turn, results in restraints from overuse of the woodland.

The same leveling mechanism also works to bring about a "socially-inclusive innovation process," in that innovation initiated by a particular individual eventually spreads to the entire community, thereby raising the standard of living of every household, notwithstanding the existing difference in labor composition and access to resources needed for innovation. As we shall discuss further below, because farm innovation is embedded in the sharing economy, trickle down works at the local level.

Chronology of innovation

Bemba livelihood activities have been changing through adaptation to the cash economy and cash crop farming, besides retaining a basic livelihood strategy with the focus on long-term food security. Along with landmark social and political events, Table 9.1 traces three major periods of innovation

Table 9.1 Three major periods of innovation in the history of Bemba agriculture

	Events	*Government Policy*	*Changes in livelihood and farming system*
1800s	Famine, out-migration		Introduction of continuous cropping
1900s	Colonization	Abolition of "traditional" *chitemene* farming	Introduction of mound cultivation
1914	First World War	Obligatory mound cultivation with sweet potatoes in semi-permanent fields	Mound cultivation in semi-permanent fields spread widely, and then disappeared after the Second World War
1920s			Mound cultivation was adopted into *chitemene* system
1930s	Creation of crop rotation system in *chitemene*		
	Labor migration	Introduction of poll tax	Penetration of cash economy
1940s			Female villagers started to sell finger millet beer brewed by a new method
	Creation of cash circulation system		
1950s			
1960s		Cultivation of maize as a cash crop	Some villages adopted maize cultivation but abandoned it when they suffered from hunger
1964	Independence		
1970s			
1980s		Encouragement of hybrid maize cultivation by providing seeds and fertilizers on credit	Some male villagers started hybrid maize cultivation on receiving a loan from the government
late 1980s			Female household heads were faced with difficulties in securing male labor, with economic disparity growing
1990s		Structural Adjustment Program	A new method of brewing beer was introduced, which gave women the economic strength to hire male workers
	Dual livelihood strategy: cash crop cultivation and *chitemene*		
2000s		Rigid enforcement of the Land Act (amended in 1995)	

Source: Field survey.

during the past 100 years: (1) the establishment of crop rotation in the 1930s; (2) the emergence of cash circulation in the 1940s; and (3) the creation of a dual livelihood strategy in the 1990s.

Establishment of the crop rotation system

The establishment of a crop rotation system can be traced back to the famine in the late nineteenth century. Villagers still remember it because most of the Bemba left their homeland and fled to the land of the neighboring Fipa people. They stayed there for one to two years and labored as farm workers; in the course of this diaspora they picked up the mound cultivation technique of the Fipa.[5] After they returned home, they modified the method to better match the local environment. Another important innovation imported from the adjacent Lungu people was to change the cultivation cycle of finger millet. Originally, the Bemba had planted finger millet only once in *chitemene* fields, but the Lungu method of planting two years consecutively spread in the 1930s, thereby reducing the risk of hunger.

After British colonization, the *chitemene* farming system was prohibited by the government as a "primitive" method and was to be replaced by permanent cultivation with mounds. In 1914, at the outbreak of the First World War, the colonial government forced each villager to create a permanent field to make 200 mounds per person for sweet potato cultivation in order to supply food to locally conscripted soldiers. As soon as the war ended, however, the villagers quit mound cultivation in permanent fields. Instead, they began to make mounds around the *chitemene* field to plant sweet potatoes. Several Bemba youths who came back from labor migration to South Africa brought new varieties of sweet potatoes, which became a favored supplementary food in the daily diet. In the 1920s and 1930s, several other crops were introduced into the *chitemene* system, for example, red finger millet from the Copperbelt region, maize from Tonga-land to the south as well as tobacco, cow peas and improved varieties of beans. By the early 1930s, mound cultivation had been built into the *chitemene* system. Cassava, which was introduced between the late 1920s and 1930s, further extended the rotation period of *chitemene* and increased the variation in their staple food repertoire. The colonial government promoted cassava production when famine occurred due to extensive crop damage caused by locusts. Cassava cultivars were also introduced by villagers through multiple informal routes. Thus, the *chitemene* crop rotation system, the first landmark innovation in Table 9.1, was established. The production of food on each *chitemene* field in terms of calories gradually increased, supporting a growing population.

Emergence of the cash circulation system

Apart from the innovation in farming technology, another most important innovation was the introduction of the sale of alcohol made from finger millet, which connected the village economy with the national cash economy. The establishment of a cash circulation system had its origin in the long-distance labor migration to the copper mines in the 1930s. Due to the poll tax introduced by the colonial government, villagers began to engage in cash-generating activities. Enhanced by the government policy of supplying labor for the copper mines, labor migration from Bemba-land to the mines became increasingly popular as a principal means to earn cash income.

These migrants came back to the village with cash as well as new methods of cash-generating activities. Wage labor and a new method of brewing local beer for sale were introduced, which expanded the use of cash in the local areas. Younger male household heads began to employ labor for cash to do agricultural work, as they had enough money to pay. By the 1940s, it had become possible for people to earn money in economic transactions within the village, which widened economic differences and changed the political power balance between the elders and the younger generation. Thus the "path of money," as it is called by the villagers, emerged within the local community. The concept denotes transactions through the medium of money and thus represents the logic of the cash economy. There is no room for "sharing" in this path. Giving money to others is thought to be payment and asking for money to be a shameful act.

Before the emergence of the path of money, there used to be only two paths: the "path of food" and the "path of respect." The difference between the paths corresponds to the manner of transaction, the people involved, and the symbolic meaning of things that are exchanged. The path of food means food sharing among relatives, neighbors and friends. Expressing one's gratitude is unnecessary or sometimes inappropriate because "food must be shared." Nobody can claim individual ownership of food; conversely, those who do not have food are free to claim a share. The people "along the path" (the people who are involved in this kind of transaction) are children, kin-group members, neighbors and close friends. Meanwhile, the path of respect concerns the obligation and politeness to others. In this method of transaction, expressing gratitude with politeness is the greatest requirement. Those who receive goods and/or services are treated as guests who are also required to be polite. Asserting a claim on something is avoided because it is regarded as shameful. This path leads to the elders, chiefs, elder members of kin-groups and guests. These two paths are essential for the leveling mechanism to function.

With the advent of the path of money, some younger women introduced a new method of brewing finger millet beer for sale, which they had acquired in the Copperbelt region.[6] Until then, the local beer used to be brewed only on the occasions of village rituals or communal labor. Two different ways of brewing finger millet beer emerged: (1) beer brewed in the "traditional" way (*chipumu*); and (2) beer brewed using the newly acquired method (*katata*). While *chipumu* was used only for communal drinking, *katata* was brewed primarily for sale.

The villagers developed their own rule regarding local beer that connected the path of money with the path of respect. According to this rule, when a woman brews beer for sale, nearly half of the raw materials should be left to be brewed in the "traditional" way and provided for free at communal drinking occasions. This did not prevent women from rising in status by hiring men who later came to buy local beer brewed by these women. In this way, money flows from cash-abundant males to cash-scarce females were introduced into the village, which contributed to stabilizing the livelihoods of women.

Creation of the dual livelihood strategy

The third major innovation was the spread of hybrid maize as a cash crop in the mid-1980s, which eventually led to the establishment of a dual livelihood strategy that combined cash crop and subsistence crop production. This process occurred nearly 40 years after the second major innovation, although the foundation for accepting maize as a cash crop was laid in the 1960s as part of government policy to introduce maize cultivation in some villages. This ultimately proved to be a failure. However, in the mid-1980s when the Zambian government tried again to boost maize production by introducing the cultivation of hybrid varieties in semi-permanent fields, villagers responded more enthusiastically because the scheme was designed in a participatory manner.

In 1985 the government added a credit system, in which villagers could get chemical fertilizers and hybrid seeds on favorable terms. This appealed to young male household heads who were eager to try new crops and techniques to earn money. Registered villagers bought hybrid maize seeds and chemical fertilizer in required quantities on credit and repaid the debt after the harvest. The policy also assured that the NAMBOARD (National Agricultural Marketing Board) would come to the nearby depot to collect the harvest at a fixed subsidized price and issue a check on the spot. Male villagers who had already made attempts to grow maize jumped at this opportunity, making great efforts to create semi-permanent maize fields.

It did not take long for other male villagers to join this credit system to start hybrid maize cultivation as an addition to the *chitemene* farming. By 1988 eight out of thirteen households (the other five households were female-headed and an elderly couple) in our study village took part in maize cultivation because it was incredibly profitable, with the amount written on the check being more than 10 times their previous yearly income. The men who commenced maize cultivation spent five to eight hours on agricultural work per day, which was four to five times longer than what they did in 1984. At the same time, female villagers maintained the *chitemene* cultivation. A dual livelihood strategy emerged based on a new division of labor with men producing hybrid maize for cash and women cultivating *chitemene* for daily food. Female-headed households were left behind and encountered difficulties in finding male labor for the cultivation of maize or even for creating their own *chitemene* for food production. How villagers overcame this problem is described in the next section.

Emergence of a socially inclusive innovation process

The introduction and the diffusion process of hybrid maize cultivation (locally called *faamu*, derived from the English word "farm") between the 1980s and early 1990s provides the best example for understanding how successful innovation occurs in the village. It happened when the Bemba incorporated new crops and techniques into their own social context and thus created a new method of cash circulation.

The advent of hybrid maize cultivation between 1985 and 1989 appears to have temporarily widened the economic disparity among households, undermining the existing system of sharing foods and other daily necessities. As has been mentioned above, a quarter to nearly half of households were headed by females in the Bemba villages at the time. They suffered from not being able to hire enough male labor to prepare *chitemene* for food production and clear maize fields for generating income. At the same time, communal labor arrangements in return for beer, or help from their close kinsmen, proved to be in vain or not enough to mitigate hunger. Another problem was that women's request to raise the price of the finger millet beer was turned down by the village headman and the chief. Moreover, women had to reduce the frequency of beer sales due to the lack of finger millet. As a result, cash circulation in the village became stagnant. The gap between rich and poor villagers became increasingly apparent.

It was in this context that innovation was born. It involved redefining maize as staple rather than cash crop. Originally, maize was considered a cash crop and in line with the ethics associated with the "path of money" it was shameful to ask for a share of the maize. As the villagers did not eat maize as their daily staple at that time, it was not shared along the "path of food." Between 1985 and 1989, female household heads encountered serious difficulty in securing male labor for clearing *chitemene.* Young female household heads abandoned their efforts to clear *chitemene* large enough for their own subsistence. Instead, they started working for wages in the hybrid maize fields of other villagers. The wage was originally paid in cash or other basic necessities such as salt and soap, which were gradually replaced by payment in maize.[7] Young women tried to eat maize as a substitute for finger millet and cassava, spending a longer time for processing maize with other women, particularly with their sisters or matrilineal cousins, whose husbands owned hybrid maize fields. Although a milling machine was available in a nearby village where the chief resided, they could not afford to pay the milling fee. As the work of pounding and milling maize is too hard to do alone, the women cooperated with each other in processing maize, and even shared maize meals together. Thus, the habit of eating maize spread among young women in the village, facilitated by the act of cooking and eating together.[8]

Yet another new form of cooperation emerged, this time between the young female household heads and elder women. This kind of cooperation was formed along matrilineal lines and uxorilocal patterns of residence, where mother-daughter relations are long lasting even after the daughter's marriage. Elder women have an advantage over their younger counterparts in securing labor. Thanks to their seniority they could obtain it through the "path of respect." Their sons, nephews and the husbands of their daughters had no option but to accommodate the elder women's requests to prepare their *chitemene* fields. These older women not only utilized this advantage but also added other methods such as asking for communal labor in return for traditional beer so they could hire male labor for their *chitemene* preparation. Younger women (mostly the elder women's daughters or their sisters' daughters) depended on the food from their mothers' *chitemene* as a daily diet supplement. The younger and the older generations cooperated not only in *chitemene* farming, but also in processing finger millet and cassava. They even shared meals, in which maize was becoming increasingly popular as the material for making their staple food, *ubwali* (hard porridge).

Thus, by the beginning of the 1990s, when maize was recognized by villagers as an important staple food, women invented new forms of cooperation among female-headed households. This change was significant in making food sharing easier within the cash economy, but it was only a first step. The economic difference between households with male labor and those without continued to widen until a breakthrough in 1991, again in the method of brewing beer.

Opening access and bottom-up diffusing

Introduction of a new *vestini* brewing method marked an epoch in the diffusion process of *faamu* cultivation. *Vestini* was a quick, easy and innovative method for young women to generate cash. It was introduced by a young female household head, who had learned how to make *vestini* in a nearby town. The merit of *vestini* lies in the fact that it needs only a small amount of finger millet and water and compared to the existing *katata*[9] require less time. Another merit of *vestini* brewing was that the brewer does not have to follow the local rule to prepare the same amount of beer for sharing as for sale. Most importantly, the women could set the price of beer themselves. The price of *vestini* beer was set at a higher level than that of existing local beer so that women could acquire enough cash to buy what their households really needed.

With the increased sales of *vestini,* cash circulation was boosted once again. The customers of *vestini* were primarily the male household heads who were rich in cash and maize. Buying local beer to offer to others was one of the social obligations for those regarded as wealthy. They frequently bought *vestini* to drink with neighbors and friends as they had done before. The money they spent on local beer went to the female-headed households and was used to employ male labor for farming as well as to buy foods and other daily necessities. Those who were rich in maize and cash tried to "share" what they had with the female household heads in the following way. They provided opportunities for other villagers to engage in "unnecessary" labor, ignoring that they already had enough manpower in their households. By using wage labor, they seem at first glance to have simply followed the "path of money." A closer look, however, indicates that it also overlaps with the "path of food" and the "path of respect" in the sense that grain-rich villagers extended assistance to the female household heads, who were given opportunities to obtain cash and maize as a reward for their work. At the same time, females got the chance to acquire the necessary skills and knowledge about the new technique of growing hybrid maize.

The introduction of *vestini* enhanced the possibility of getting necessary male labor for hybrid maize cultivation. In particular, the unmarried young males, who were once prohibited from drinking beer, started drinking *vestini* in return for their labor in clearing maize fields and other work. The young female household heads were successful in preparing their own maize fields for cash crop production as well as securing the labor needed for cutting their *chitemene* fields. Furthermore, the aforementioned cooperation between elder and younger women evolved into a new form of "division of labor" between mostly mothers and daughters in which the younger female household heads were chiefly engaged in *vestini* sales to get cash, while the elders concentrated on the *chitemene* cultivation for food production. The daughters, for their part, supplemented their food supplies with vegetables from their mother's *chitemene* fields, while they asked male customers who drank on credit to create *faamu* fields for their mothers. In this way, both younger and elder women found a way to have their own maize fields as well as the *chitemene* fields.

Another important aspect of the above process is that maize has now become an important staple food. As women earned their reward for the work in the maize fields, eating maize as their staple food became a common practice, first as a substitute food and later as one of their staples, as noted above. For women, the dissemination of maize was a game changer. It provided material for *ubwali,* even during the rainy season, when they normally depended on cassava. Equally important, it liberated them from the routine work of pounding and hand milling. With the increased cash income, they could afford to mill the maize using machines in the nearby town, which had not been possible in the early 1990s.

As soon as maize was evaluated as "food," it was included in the local sharing system, where it is acceptable that those who are lacking in food can ask someone else for maize. As maize was no longer just a cash-equivalent item, it became connected to the "path of food" and something that could and should be shared with other villagers. It became embedded into the local sharing system where long-term food security and stability of livelihood are expected. Maize can now be exchanged along either the path of money or the path of food, according to the context. Thanks to this conversion system, which bridges the cash economy and communal relations, villagers have been able to incorporate cash into their social system.

Modelling local innovation history

Using the three cases of innovation discussed above, particularly the 1980s hybrid maize diffusion story, the process of innovation can be modelled as a set of five stages, as shown in Figure 9.1.

In the first stage of the process, each villager engages in multiple acts of innovative trials. Some of the trials are derived from experiences with past

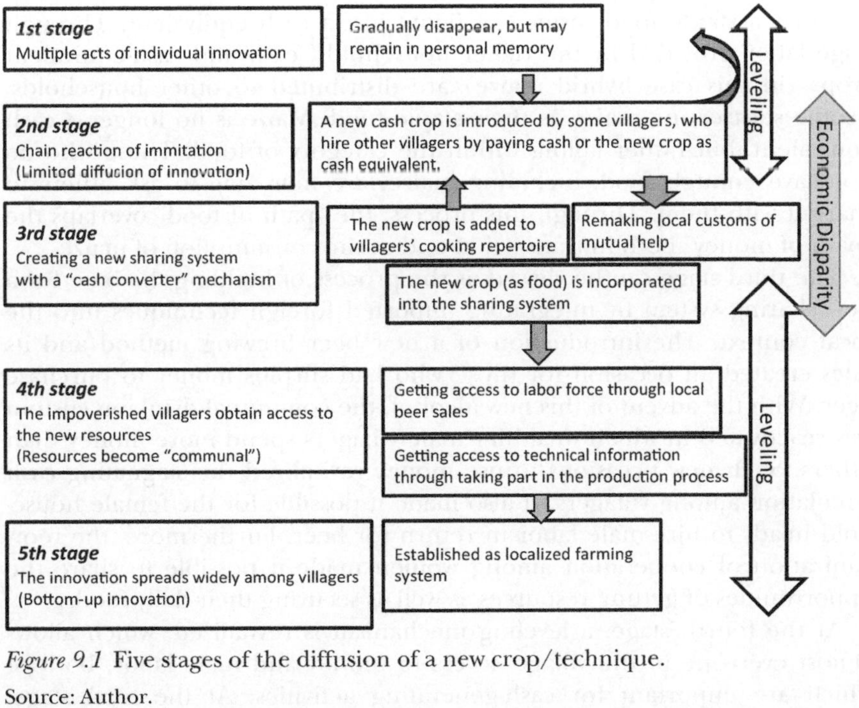

Figure 9.1 Five stages of the diffusion of a new crop/technique.

Source: Author.

government projects or those of foreign non-governmental organizations (NGOs), which were not successful at that time. It is notable that the original sources of new techniques vary according to each villager, and several different experiments may be attempted simultaneously by the same person. If other villagers show no interest in these trials, they will most often come to nothing. A few, however, may be remembered and could be utilized at a later point. Once other people confirm that an innovative trial of a pioneer is successful, several villagers, especially those who have the necessary resources, follow suit, which in turn may trigger a chain reaction of imitation among villagers. Basically, a new crop is adopted provided it meets the standard of enhancing their food security. At the first and second stages, the leveling mechanism based on the ethic of sharing serves to restrain economic differentiation.

At the second and third stages, however, the economic difference expands between those who have the resources and those who do not, as the existing rule of sharing does not cover money and monetary resources. The new crop is still recognized as a cash-equivalent to be exchanged within the "path of money." Cash circulation in the village eventually stagnates until a new sharing system is created through the invention of a system converting cash to other sharable goods.

In the transition from the third to fourth stages, the expanded economic disparity is gradually leveled among the households due to the evolution of a new manner of sharing. The newly introduced cash crop becomes a staple food instead of being just a cash equivalent. Through wage labor provided by the richer households, cash and cash-equivalent crops (in this case hybrid maize) are distributed to other households. People start eating maize as their staple food. Maize is no longer a cash equivalent, but rather falling within the category of food. Those who do not have enough food, including maize, are now able to ask others to share it with them. Through this process, the "path of food" overlaps the "path of money" both in transaction and in the consumption of maize.

The third stage can be viewed as the process of building the base for a new sharing system by integrating imported foreign techniques into the local context. The introduction of a new beer brewing method and its sales created an occasion for those who had surplus money to purchase beer. With the advent of this new alcohol, the communal drinking custom was reactivated in which wealthier male villagers spend more money than others. Such new occasions to use money revitalized the stagnating cash circulation among villagers. It also made it possible for the female household heads to hire male labor in return for beer. Furthermore, the reorganization of cooperation among women made it possible to share the opportunities of getting resources as well as securing their daily food.

At the fourth stage, a leveling mechanism is revitalized, which allows almost everyone in the village access to money and male labor, both of which are important for cash-generating activities. At the final stage, innovation widely spreads across the whole community, and "bottom-up innovation" is realized by which everyone benefits from the innovation and can improve their standard of living. After several years of improvement, new crops and skills may be modified to become part of the local technology.

This model may also be viewed as a circular loop with its own critical junctures. One such juncture is in the first stage when villagers assess the value of a newly introduced crop. Another juncture is in the transition from the second to third stage when the maize crop becomes the daily food to be shared with others. This transition from being a cash crop to becoming a staple is initiated by the better-off villagers who have easiest access to the necessary resources for pursuing innovation but once the crop is being treated as food rather than a cash equivalent, this transition process is also promoted by those who lack necessary resources to secure their livelihood. The local ethic of sharing kicks in and innovation spreads to all villagers. Communal management of the resource is gradually built through the daily practices and interactions of people.

Policy implications of the Bemba case

How the Bemba people pursue their livelihood is not the only case even if it inevitably has its own features. As we have shown in this volume and as summarized in Chapter 8, there are important policy implications to be drawn from a closer look at how agriculture fits into the way most Africans make their living. The conclusions that we draw from the Bemba case, therefore, should have wider application in the African context. Drawing on our own studies, we wish to highlight the following three implications that deserve greater and continued attention not just in research but also policy circles: (1) technical innovation is socially embedded; (2) innovation is place-based; and (3) innovation involves new combinations.

Technical innovation is socially embedded

In his attempt to theorize the innovation and diffusion process of Japanese agricultural technology, Soda (2006) argues that social aspects are as important as economic ones in the introduction of a new farm technology. He classifies the behavioral patterns of Japanese farmers to innovation into three types: (1) pioneers, (2) adopters and (3) imitators.[10] Although the pioneers are very few, once they carry out a trial for innovation, some adopters follow suit if they find it has merit. The innovative technology is widely diffused when the imitators adopt it. In the process of Japanese farm innovation, the pioneers tend to be the leaders or men of repute (and often of wealth) in the village community.

Similar processes of diffusion of a given technology are found among the Bemba. However, the division of roles between pioneers, adopters and imitators is not as clear because they are interchangeable. Generally, Bemba villagers are so curious about new crops and techniques that everyone is ready to try them. Even a poor elderly woman may try growing a new variety of crops if it contains an element of appeal. If others do the same, it sets in motion a chain reaction that ensures innovation will be diffused across the community quite quickly. Cash circulation has played an important role both to connect villagers and to promote innovation. The "path of money" provides an opportunity for villagers to create intimate, face-to-face relations with strangers, which also gives them an option to extend their social networks. Furthermore, this path has enabled them to incorporate resources equivalent to cash into the daily circulation of not-for-profit transactions. The creation of this conversion channel made it possible to connect the local economy of sharing with the national/global cash economy. At the same time, the innovation processes of the *chitemene* system and the lifestyles of the Bemba villagers show that their livelihood strategies have constantly followed the principle of securing food and stability of life amid ever-changing socioeconomic circumstances. Even though the national cash economy has penetrated local life, people have eventually invented ways to articulate the local exchange system partially to the national cash economy. They have decided to stand with one foot in each system.

Innovation is place-based

Since the publication of Paul Richards's formative book *Indigenous Agricultural Revolution* (Richards 1985), numerous researchers have explored the advantages for agricultural development of farming systems based on local knowledge. In his book, Richards presents the African peasants as actors who have adapted themselves to the specific local environment through constant innovation of farming techniques based on keen observation of their surroundings. He provides a clear perspective on local peasant knowledge and practice and characterizes it as "peoples' science." By doing so, he criticizes the many policymakers and development researchers who regard the local way of farming among African peasants not only as an obstacle to progress but also as the cause of deforestation, environmental degradation and underdevelopment in general. In a similar vein but with a wider scope, Shigeta and Kaneko (2017) examine the concept of local knowledge (*zairai-chi* in Japanese)[11] as an interactive process between people and the local environment, including their sociopolitical condition.

More recently, local people, researchers and policymakers alike have attempted to create an alternative future design to acknowledge and realize the place-based sustainable lifestyles in rural Africa. As this chapter shows, the local farming system should be treated as a socio-ecological system of the utilization of the environment, and therefore, agricultural innovation should be treated as a "social process." By tracing the history of local innovation, it is possible to identify clues to the potential for place-based, sustainable innovation of local farming systems and rural livelihoods. Rethinking African agriculture from the perspective of local innovation histories is meaningful, because it forces us to reconsider the role that the community plays in innovation, which in turn gives us an indication of how to design an alternative model of development.

Innovation involves new combinations

Schumpeter's classic argument on innovation broadly encompasses not only technological change in a narrow sense but also the introduction of a new method in economic activities (Schumpeter 1981 [1934]). His famous definition of innovation as the "new combination (*neue Kombination*) of different elements" can also be applied to the Bemba – but with significant modification: Schumpeter's "new combination" occurs mainly in the sphere of production while African peasant innovation is based on a new combination of both productive and consumptive activities. As African villagers bridge these two different spheres, they reinvent social relations and norms. As we have seen in the case of the Bemba, innovation succeeds when it is in line with villagers' strategy to secure the long-term stability of their livelihood but at the same time is part of the cash economy, where existing social networks and the sharing system within communities are updated to be reconciled with the contemporary context. Thus, while they are aware of the advantages of modern technologies in the context of a cash economy, they are not mindless followers of the agricultural modernization policies of government. Rather, they attempt to discern which technique is suitable for their livelihood pattern. Such a process is based on copious knowledge of the woodland, which has been accumulated through long interaction between the environment and the practices of the villagers.

Conclusions

It has become increasingly clear in international development circles that for innovation and reform to occur they must be grounded in a local context and be driven by local "champions" or pioneers. Putting the thesis into practice, however, has proved quite difficult. We have provided insights into how it can be done. Among the markers we have laid down are: (1) the importance of providing a historical perspective on change in order to identify the local dynamics that produce it; (2) the necessity of acknowledging that development in rural Africa occurs in communities of consumption rather than production; and (3) individual and incessant trials of each villager in search of personal gain eventually enlarge opportunities for a wider community to make innovation socially inclusive.

These are the opposite of what has characterized official policy toward Africa for several generations; Africa has instead relied on a top-down approach focused on raising productivity only and assumed that African villagers are backward. That is why this story of how local innovation takes place deserves to be further discussed in its broader international development context.

Notes

1 Some of the results of our research have already been published in Sugiyama (1987, 1992), Kakeya and Sugiyama (1985, 1987), and Kakeya et al. (2006).
2 *Miombo* is a local term denoting tree species belonging to the genus *Brachystegia*.
3 The regeneration pattern of the *miombo* woodland of this area has been modified to correspond to the Bemba lifecycle through everyday livelihood activities. Repetitious cutting of tree branches makes the tree shapes suitable for *chitemene* work: the trees become lower in height and grow multiple branches with thick leaves.
4 However, the greater number of male members does not necessarily result in much larger farmland, because female labor is indispensable in carrying the branches to make *chitemene* fields.
5 Interviews with the village elders. Audrey Richards's field notes of Bemba agriculture in 1931 (Richards 1931) recorded a similar account.
6 According to the elder women, they learned the brewing method from the Tonga people who were living together in the cities. This corresponds to Colson's (1988) description.
7 The villagers regarded this type of labor employment as "sharing" the opportunity for the have-nots to get maize and money, as we shall see further below.
8 Richards (1939) observed similar "cooking groups" in the 1930s, which functioned to create a new cooperation among female villagers.
9 In *katata* brewing, women have to begin with making the natural yeast from the sprouts of finger millet, which takes more than two weeks. Amounts of 200 liters of water and 40 kg of finger millet sprout powder are required for brewing. On the other hand, the *vestini* brewing method takes only a few days. It requires 20 to 40 liters of water, 4 to 6 kg of finger millet sprout powder, and 500 Kwacha (equivalent to approximately US$1.5 in 1994) for sugar and dry yeast fungus powder.
10 Although Soda refers to Schumpeter's economic development theory (1934), he does not adopt Schumpeter's term "entrepreneur" and uses "pioneer" instead in order to stress the farmer's creativity (Soda 2006: 200–1).
11 They carefully choose the term "local" instead of "indigenous," in order to avoid the nuance of "unchanged," and to stress that people's knowledge includes flexible change and constant improvement.

10 General conclusions

Implications for research and policy

Goran Hyden, Kazuhiko Sugimura
Tadasu Tsuruta

Introduction

The purpose of this volume has been to contribute to a fuller understanding of agriculture in Africa by going beyond the often ill-conceived scientific inputs that have been the mainstay of how the international community, in collaboration with African governments, has approached the task of raising rural income and overcoming poverty. While science obviously has an important role to play in African development, the rather indiscriminate manner, in which it has been applied to help accelerate rural development, has often amounted to no more than shots in the dark. In virtually all the literature on development in Africa, not just its rural areas, the continent is treated as a mega-project that will be realized through adherence to policies and strategies developed far away from the social reality on the ground. Thus, Africa has typically been identified as a region in need of extra attention. It started with the World Bank issuing special reports on Africa (World Bank 1981, 1989) and, although Africa, according to OECD figures, has been receiving roughly the same share as Asia of the total Official Development Assistance (ODA) – some 35 percent – it is Africa that has been identified as the most worrisome case. The premise that development challenges in Africa are especially serious is not mistaken but the reasons usually advanced such as faulty policies or unreliable institutions – both related to a weak science-base – are not adequate explanations. A search for understanding Africa's predicament must go further than official policy and formal institutions. Scientific solutions must be problematized in a historical context, not just in terms of what westerners see as theirs but in the broader Big History perspective that explains why Africa is different and why western science so often misses its targets there.

The contribution by the individual Japanese researchers featured in this volume and the intellectual tradition – legacy of the Imanishi School at Kyoto University – on which they build their work, opens new vistas on African social realities that have at best received only marginal attention in the scholarship of European and American academics. The Japanese perspective, as this volume highlights, is especially relevant for understanding the problems associated with agricultural development but it has wider applications in the study of both economics and politics. In this final chapter we will summarize what we see as the contribution that this volume makes to our knowledge of Africa by focusing on (1) the importance of Big History, (2) the main insights from the case studies, (3) their policy implications, and (4) where Africa may be heading given its predicament.

The relevance of Big History

The manner in which we refer to Big History in this volume has some similarities with Fernand Braudel's *Annales* approach which recognizes the slow and often imperceptible effects that factors in the human environment such as climate, land and technology have on the evolution of society. His scholarship and that of his disciples went no further however, than exploring specific countries like France, regions like the Mediterranean, or civilizations like western. While this approach was a great step forward from the previous emphasis on the shorter term and, in particular, the study of history as a series of events, it saw the world from a purely western perspective, leaving places like Africa as being of interest only in terms of its contribution to the growth of western civilization (e.g., Braudel 1993). The Japanese approach to Big History extends to humanity at large and identifies the key stages through which human society has evolved beginning with Natural Society, an essentially stateless condition in which humans take advantage of nature but do nothing to conquer or destroy it; Agrarian Society, where a social stratification is in existence and the institution of the state is used to exploit nature and subordinate human activities to its demands; and Industrial Society, which refers to modern society and its structuration based on economically rational calculations.

The point that the Japanese scholars make, as articulated in this volume, is that African countries have yet to undergo an Agrarian Revolution that subordinates agricultural production to the commands of the state. Their social formations do not reflect the consequences of such a revolution. Instead, African societies are organized with a view to prioritizing consumption and social reproduction. Despite all the blame that academics and civil activists place on how their countries are being governed, the key issue is not that there is a state that steers the direction of society. The colonial administrators did their best to transform their territories into Agrarian Societies, but they never succeeded. Once African leaders took over at independence, they abandoned the project altogether and allowed their countries to revert to the logic of a Natural Society, according to which communities based on consumption take over public institutions with the view to sharing resources among themselves. Democratic elections have become a way of regulating the competition among such communities and it is reflected in the Ghanaian saying that "it is our time to chop" expressed by leaders of the political opposition who believe that it is their time to be in charge of government and benefit from it. The wearing away of public resources is also evident in the prevalence of corruption which has been treated by donors as evidence of personal deviation rather than as the structural issue it is, where Natural Society logic collides with the rationale of institutions derived from an Industrial Society setting.

It is necessary, as we have stressed throughout this volume, to make a distinction between agriculture and Agrarian Society. The former can exist without the latter but agriculture in a Natural Society context is pursued without the structures in place that promote a social transformation. The neglect of this point is a major reason why donors have failed in their efforts to promote peasant agriculture as a means of overcoming rural poverty. As we have shown in this volume, African cultivators are not specialists but generalists in the sense that they prefer diversification over specialization. When it comes to securing a livelihood, non-agrarian factors are often as important as tilling the land. The latter is mainly important as a means of securing subsistence and sharing food among kin and friends. When price on cash crops declines, peasants tend to plant horticultural products, typically not for sale but for household consumption. That is why horticultural exports in African countries come from specialized farms that are managed on a commercial basis by urban-based businessmen.

Participation in the cash economy is important for African peasants but not so much to enhance their income from land cultivation but rather to have enough to feed the household and to spend on sharing with others in order to meet expected social obligations. The rapid spread in recent years of microcredit and savings institutions is evidence of how people, especially women, participate in the cash economy in order to have the savings to share in the case of bereavement or any other unexpected cause. These institutions flourish in Natural Society contexts because livelihoods are little or not at all connected to demands from the state.

The relative autonomy of the peasantry posed a serious challenge to development in post-revolutionary Russia in the early 1920s. An impatient Lenin, realizing the difficulties of uprooting Russian peasants from their pre-capitalist forms of livelihoods, decided to cut this Gordian knot by socializing all land and turning the peasants into workers. Western donors have more recently used a more benign approach to address this issue but the Agrarian Question – as the Russians labelled this development challenge – is still very much a live puzzle to tackle in Africa. Neither socialist nor capitalist forms of agriculture have provided convincing answers largely because the analysis has been situated in the sphere of production and assumed that farmers are specialists. In African countries, social formations continue to be shaped by sharing within and among communities of consumption which gives them many features that do not fall within mainstream frames of analysis. As this volume has demonstrated, Natural Society legacies can be found in all kinds of social context in African countries making them different from those societies that have gone through an Agrarian Revolution.

Insights gained

The body of knowledge presented here adds insights into African rural livelihoods that have largely been overshadowed by the predominant western view of development as the pursuit of rationally designed policies aimed at addressing society as if everybody would benefit in the same way regardless of background. Macro- rather than micro-perspectives have dominated; uniformity rather than diversity has shaped the outlook. Our findings are a reminder that much of what happens in Africa takes place outside the realm of official policy yet serves as prime determinant of how people live and work. These findings add to our common knowledge in at least the following respects: (1) agricultural innovations are complex processes involving both trial and error; (2) these innovations are place-based but can spread rapidly without outside help; (3) the subsistence ethic survives even in commercialized environments; (4) the economy of sharing moves to urban areas as rural households diversify their sources of income; (5) institutions are living mechanisms; and (6) uncertainty creates its own social logic.

The complexity of innovation

The mainstream approach to agricultural development in Africa – as elsewhere – has been based on the idea that peasant farmers should learn from the "experts." Innovation has been pursued by linking peasants with governmental extension services or with private commercial farm enterprises on a contract basis. The extension services are based on the findings derived from government researchers who have the responsibility to develop new crop varieties that are more productive. Commercial farmers offer a more "hands-on" approach but are tied to the same objective of enhancing agricultural productivity among peasant farmers. Both these approaches suffer from reliance on blueprints and the assumption that change such as a sustainable rise in productivity can be created "here-and-now" within the unrealistic timeframe of specific project interventions.

The research presented here has shown not only the complexity of rural livelihoods but also the complexity of changing them. Despite living in conditions that are strenuous to manage, rural Africans are not afraid of trying new things. Any risk aversion that they have is tied to not being able to engage in change without a sense of doing it on familiar terms. Adoption does not necessarily come at once but is the result of trials that eventually produce positive outcomes and are adopted, initially by pioneers and later by other community members. What our research underlines is the importance of understanding the local history of innovation. As Sugiyama illustrates in Chapter 9 with her case study of the Bemba people, there is a local innovation dynamic that involves not only technical considerations but also how it fits into local communities of consumption. Indigenous knowledge matters, as Richards (1985) emphasized. What we add here is a better understanding of the underlying conditions of Natural Society and the process of innovation that starts from a consumption rather than production point of view.

The place-based nature of innovation

The scientific or technocratic approach to change is usually summarized in a model or blueprint that can be applied regardless of place – and time. Neither geography nor history really matters in this approach. As this volume has demonstrated, this way of developing African agriculture easily backfires because of local conditions on the ground. Peasant farmers have difficulty relating to recommendations that they cannot interpret from their own subsistence perspective. This is the reason why these peasant cultivators are often portrayed as "backward." This is a stereotype that misses the point about what it means to take on innovation from a consumption and social reproduction perspective. We have shown that these cultivators are open to trying something new. This applies not only to the already well-to-do households but even to an old woman heading her own household. The point is that a new idea may spread quite fast once someone with the ability to see the value from a local farming system perspective has decided to try it out. It is understanding the conditions of each place or community that serves as driver of change. This is a confirmation of the findings in Leonard's seminal work on the agricultural extension service in Kenya (Leonard 1977). He showed that as policy content was communicated down the hierarchy of the Ministry of Agriculture, the advice that these officers provided was intentionally adjusted to appeal to the local villagers. The reason for this conversion of policy intent was that these low-level field officers lived among the villagers and shared the same livelihood system of those they had the responsibility to serve. They did not work with everyone but chose those who showed greatest interest in adopting their advice. What Leonard demonstrates, therefore, is that informal networks between field officers and villagers translated official policy into something that aligned with the villagers' notion of what was appropriate for improving their livelihoods.

Local ethic survives commercialization

As our research confirms, African peasants do not specialize their knowledge of the land with a view to raising its productivity. As generalist managers of their respective households they regard production on the land as one of many ways of securing enough to share with all. Unlike the contemporary situation in other regions of the world, households do not operate as nuclear and autonomous units but are linked in communities of consumption that stretch across families to include related kin and sometimes their friends and neighbors. At times of emergency like famines, the readiness to share is automatically extended to everyone in need with the wealthier households taking the lead. As we have seen in Chapter 7, even if peasant farmers are relocated in a settlement scheme where they are expected to operate on a commercial basis, they take with them the norms and values that prevail in their home area. Especially fascinating in this context is how the value of a crop changes once villagers define it as food rather than cash and how social behavior changes accordingly. These are insights that rarely feature in mainstream development research yet are crucial for understanding how local economies operate and may be changed from within.

Natural Society values transcend geographical boundaries

As households have diversified to include members migrating to urban areas in search of employment, the original Natural Society ethic nowadays encompasses members spread over an extensive geographical area. In some cases, it involves household members living in the diaspora. Thus, the economy of sharing – or the "economy of affection" – is still an institution that shapes social relations and structures in Africa. It is not a one-way traffic from urban to rural areas or from rich to poor families. It involves two-way exchanges when local cultural norms call for it, for example, at key social occasions like child baptism, weddings or funerals, as has been noted also by other scholars interested in the role of remittances, e.g., Anderson-Djurfeldt (2015), who shows that food originating in the villages is shared with urban-based members of the community. The fact that households united in such communities can fall back on several possible sources of raising money or sharing food is an insurance mechanism that nowadays works not only in marginal rural contexts but also in urban settings where life in the informal sector has many parallels with the uncertainty of village livelihoods. These exchanges across wide geographic territory have been facilitated by the creation of money transfers by mobile means such as M-Pesa in East Africa. This technical innovation is likely to sustain and possibly strengthen the inclusivity of Africa's many consumption communities.

Institutions are living mechanisms

Institution has been a key concept in the international development literature for a long time. It has been used in structural terms, indicating that institutions are rules that independently shape and guide human behavior.

While this approach may explain much, for example, national variation in capitalism and the welfare state, counterintuitive paths of political development, policy variation within nation-states, and why some societies grow while others stagnate, this notion of institutions, as Berk and Galvan (2009) argue, is less helpful when it comes to explaining their own transformation, the nature of informal politics and economy, and entrepreneurship. Understanding innovation in a context of informal institutional arrangements like those we have discussed in this volume, therefore, calls for a different approach. Berk and Galvan propose an "experiental" approach which acknowledges that living under rules implies living through them: not just playing by the rule but actually playing them in the sense of improvising. Such improvisation may be blocked and turn into routine, but this is not because of rules preempting it. Instead, it is because human intervention inside institutional structures cause it. Institutional stability, therefore, is not the default position. As we have seen in this volume, peasants are constantly engaged in adapting their behavior to meet the challenges they face as their environment changes or offers unexpected opportunities. While they share the objectives inherent in their subsistence ethic, they pragmatically manage their livelihoods, e.g., by adjusting the social boundaries of their sharing or adopting a technical innovation with implications for their social relations. Achieving or restoring order, therefore, takes creative and consistent action – a way of looking at institutions that is relevant for understanding how to approach public sector reforms "from within." Institutions are not autonomous of human agency but rather shaped by prior human experience and thus projective and dynamic. This "rules-in-use" perspective resembles the view advanced by Lindblom (1990) that people do not have preferences prior to action but rather discover and revise them through action.

Uncertainty creates its own social logic

Many African peasants live in an uncertain environment characterized by unreliable weather, the presence of wild animals that consume or destroy crops, as well as increasingly facing strangers whose background and intention they do not know. While African societies have traditionally been quite effective in integrating strangers, such assimilation has become a greater challenge as the intensity and extensity of migration has gathered momentum in recent decades. The uncertainty that Africans – rural as well as urban – encounter today is both environmental and social in nature. It encourages its own social logic that the chapters in this volume have highlighted. It is based on three principles that recur in the stories shared here. The first is *diversification.* Africans on the land do not put "all eggs in one basket." They operate on the assumption that things can go wrong, whether it is crop failure or any other mishap that calls for resources from other sources than the land. We have seen how they combine farming with such activities as foraging in the forest, wage labor and temporary income-gathering from own-initiated pursuits. This generalist approach to securing a livelihood is also prompted by the social obligation of sharing resources. Unlike the farm households that Chayanov described in pre-revolutionary Russia a hundred years ago, African households are typically not nuclear but part of a wider community of consumption within which *sharing* is an obligation that is costly to ignore. We have shown in this volume that this logic prevails even as villagers move into settings where a commercial logic is expected. This second principle, therefore, is a powerful presence in social formations in rural Africa and it contrasts, as we have shown, with the situation in Asia where the Green Revolution has not only transformed agriculture but also – thanks to a successful state-led intervention – laid the foundation for an Industrial Society with its emphasis on specialization and the decline in traditional forms of sharing. In Africa, it remains an integral part of the subsistence ethos that nobody should go hungry. Not everything, however, is being shared. Food crops are less likely to be shared than meat, whether slaughtered cattle or wild meat killed in the forest. The third principle, therefore, is the preference for *immediacy*, i.e., the consumption of items that need to be eaten before they go bad. Grains are typically not shared, the exception being times of famine, but it is interesting how the identification of a specific grain as use for subsistence changes the local perception away from commercializing it for the sake of earning more money.

Policy implications

The western notion of the state is that it is an instrument in the hands of citizens for the purpose of achieving common objectives. It is a rational creation that has grown out of the experiences of going through initially an Agrarian and later Industrial Revolution. It found its final form in the confluence between nationalist and democratic ideals in the beginning of the twentieth century. It became the nation-state that scholars tend to take for granted. Africa, however, does not have nation-states but state-nations. In colonial days, people living in their own communities were brought together into sovereign territories where the state was unquestionably the key institution. Among other things, it became the basis for a notion of citizenship in African countries and, by extension, the emergence of civil and political rights. By denying these rights to the majority population, the colonial masters gradually found the educated elite among the Africans claiming the same freedoms as applied to the population of the metropolitan countries in Europe. This eventually led to political independence for the African colonies. When the Africans took over, they inherited a social order in which the primary role of the state was to integrate a multitude of ethnically based communities into a nation. This challenge is still ongoing which means that the state in African countries remains foremost an instrument for retaining social and political order rather than producing change and development. This also means that there is a disjuncture between what people want and what the state does. In a Natural Society setting where communities pursue their livelihoods quite independently, the state is experienced primarily as "policeman." That is why donor-funded development projects rarely fit very well into how government institutions operate. Being highly bureaucratic they lack the sensitivity and flexibility needed to address Africa's complex forms of social organization. It is far from being a Hobbesian state of nature and rather a very rich and varying network of communities that have in common the ambition to organize livelihoods that combine social inclusivity with environmental management care. Non-governmental organizations (NGOs) have greater flexibility than government bureaucracies, but they are often so tightly linked to a western ideology that limits their ability to change African communities from within. The question, therefore, is how outsiders and governments on the continent can adjust their approach in ways that better relate to the circumstances on the ground.

This takes on added importance today in the light of growing calls by African countries to better control their own destiny, and a realization in donor circles that they must "work with the grain," as Levy (2013) calls it. This means building on local initiatives rather than replacing them with external models of policy action. Putting together an approach that makes sense and stands a chance of getting attention in policy circles would entail a three-pronged set of measures aimed at advancing local ownership and growth of indigenous initiatives: (1) setbacks are an essential ingredient of progress; (2) politics rather than rationality determines policy outcomes; and (3) an enabling environment provides a better opportunity for learning and sustained action than compliance with a specific project or program blueprint.

Learning from setbacks

Defining and measuring development has largely been the prerogative of economists who, thanks to their dominant influence over the way we perceive development in largely linear terms, have set the parameters of the main discourse. This notion of development as running a race or climbing a ladder may have its advantage of demonstrating performance, but it also has a serious drawback by ignoring attention to how success becomes sustainable. If all energy is spent on reaching a goal at whatever cost, there is little space for learning what is happening and why actors end up with different levels of achievement. One of the few economists that have challenged this orthodoxy is Albert Hirschman who in the late 1950s suggested that getting things right depends on first having experienced and acknowledged the ways they can go wrong. Unlike most of his fellow economists, Hirschman recognized the value of disequilibria. In his study of how nations in Latin America were making progress using different approaches to deal with critical policy problems, he criticized the overconfident belief that all problems were inherently solvable and questioned the then prevalent notion that reform could be achieved only by opposite processes of violent revolution or of peaceful change. Hirschman argued that both routes in fact entailed shifts of power and wealth, and that such shifts were produced by contriving alliances engaged in what he called "reform-mongering," a method of action that used unsuspected and unorthodox opportunities for maneuver and advance (Hirschman 1963). While being at heart an optimist his foremost contribution was to highlight the need for a good dose of doubt in adopting specific policy solutions. He was opposed to the notion that there is only one way to do so.

Hirschman suggested that good things eventually would come from what he viewed as the constructive tensions between private interest and civic-mindedness and between quiet compliance and loud protest (Hirschman 1970). His reasoning was based on what he observed in the Latin American economies where urbanization and industrialization had already produced a civic public realm (or were in the process of successfully doing so). Where government institutions function in an environment characterized by the prevalence of Natural Society values and norms, the context is more ambiguous and complicated. Political parties are not anchored in specific policy positions and politics is not a competition of ideas but rather a rivalry centered on access to consume public resources for what is tantamount to "club goods," i.e., outcomes that benefit specific groups rather than what renders government truly public. This dynamic does not produce constructive tensions between private interests and civic-mindedness or between quiet compliance and loud protests. Instead, policy tends to be formulated in a "vacuum" often guided by donor funding and calls for compliance with international accords like Agenda 2030 and its Sustainable Development Goals (SDGs). Donors and governments in Africa have a lot to learn from the many setbacks they have suffered.

Limits to rationalism

Development policy analysts have had an in-built preference for acting "big." The assumption has been that development in Africa involves a social transformation triggered by concerted political action. This was evident right from the beginning in the early 1960s when development planning was adopted as the principal instrument for generating a rational and comprehensive approach to policy. Because there was a shortage (if not total lack) of local economists trained to take part in this revolutionary venture, the burden of putting together these ambitious planning documents fell on the shoulders of economic scientific expertise hired from outside Africa. Planning in those days was not a "socialist" cause but was equally embraced by liberal economists. Thus, planning ministries (or commissions, as they were called in some countries) were dominated by expatriate advisors from the United States, United Kingdom, the Nordic countries or socialist economies like Poland and Hungary. They exercised a strong influence over how these national blueprints were authored and thus how various policy priorities were aggregated into a single document. While national development plans became a way for donors to justify spending in African countries, they never really became guides for local policy action.

The limits of rationalism in policymaking had already been observed and recognized in studies of policymaking in democracies like the U.S. Challenges to the rationalist orthodoxy are associated with two names: Herbert A. Simon and Charles E. Lindblom. These two economists straddled disciplinary boundaries and had perhaps their greatest influence in the fields of public administration and political science.

Simon (1957) coined the notion of bounded rationality in his study of how decisions tend to be made in organizations. He drew attention to three factors that limited the reach for optimal solutions based on a systematic search followed by a selection of a preferred choice based on cost-benefit calculations. The first of these was the intractability of most problems. The analyst or decision-maker typically encounters the fact that the search for solutions has no clearly cut endpoint. The inclination, therefore, is to bring the search to an end once there is a sense that the process has led to a satisfactory answer. Simon, therefore, sets optimization in contrast to his own alternative "satisficing" approach. Another factor is the cognitive limitations that decision-makers face. Even when assisted by the best brains, policy problems tend to be bigger than a group of policy practitioners can handle. This means that potentially valid options are ignored simply because the human mind cannot perceive them. Wittingly or unwittingly, policymakers choose suboptimal options because their grasp of the issues is not wide or deep enough. The third factor is the time available for making decisions. The time frame for any decision tends to be much shorter than is necessary for a comprehensive search for and computation of an optimal policy option. In sum, the pursuit of perfect rationality is a waste of time and unrealistic. Simon's version retains the faith in rationalism but accepts that it needs to be applied with a sense of realism and pragmatism.

The same argument is reiterated in a somewhat different context by Lindblom (1959) who maintains that a series of small steps, even when out of sync, are more effective than attempting a single great leap forward. He was questioning the usefulness of striving for a comprehensive understanding of the problem prior to action and developed his own way of analyzing policymaking by focusing on the process, i.e., how decisions are made in political settings. Policy processes are evolutionary which, according to Lindblom, means that decision-makers do not start from scratch but operate with already existing knowledge on which they try to add whatever else makes sense in the context of making a final decision. Policy analysis is inevitably ad hoc and often disjointed and like Simon, he accepts that policymakers look for quick support for their decisions rather than engaging in a time-consuming comprehensive review of facts. His "disjointed incrementalism" was criticized by many development policy analysts for ignoring the need for an imaginary and visionary approach to solving the big problems facing developing countries. After all, Lindblom's analysis was based on what happens in the United States where policy does not change in leaps and bounds. His point, therefore, is that public administrators and policy analysts prefer incrementalism not only because they are desperate in trying to simplify their problems but also because they want to be relevant.

This debate about how to solve policy problems that took place over 50 years ago is still relevant in the African context because the tensions between acting "big" and effective implementation have continued to be a prominent feature of the policy process. Fundamental decisions are made without adequate respect for what it means to put them into practice. The study by Therkildsen (1988) on donor-funded rural water supplies in East and Southern Africa showed that the rational "blueprint" approach to implementation overlooked the need for adaptive and incremental types of decisions taken to achieve the overall goals of these projects. Donors seemed uninterested in abandoning rationalism and, as the author put it, continued to "water white elephants." In a follow-up of Therkildsen's research 25 years later, another study showed that the "watering" of unrealistically designed projects continues (Booth 2015). The perpetuation of projects that lead nowhere has a lot to do with the inability of state institutions to connect with local communities on terms that are not threatening to their members. State agents, like donors, assume that they have the answers to society's problems, which means there is little or no room for people to improvise on their own without the risk of being punished.

The disconnect between policy and citizen perspectives on how to achieve improvements in their livelihood is not only a question of communities being difficult to incorporate into the policy process. It is equally much a problem with the policy design. The various models of decision-making in use are not designed to accommodate a bottom-up approach or one that facilitates "working with the grain." The logic which prevails in African society, as suggested above, runs contrary to the basic principles of bureaucratic policymaking and it may be only through a genuine devolution of power that the social logic in society can be incorporated into some form of public policymaking.

An enabling environment

The third issue that needs to be considered to make policy more locally relevant calls into question two ideas that have been deeply ingrained in the international donor community: (1) only governments produce *national* development, and (2) development must be coordinated and coherent to be effective. These ideas have resurfaced in various versions since independence when they were officially blessed by the OECD. It has received support from African governments which in doing so have been assured of revenue for their development budget. The inadequacy of the state apparatus to reach out and put ambitious redistributive policies into practice, however, has over the years also led to broad-based public disillusionment with government, culminating in the 1990s with the rise of civil society organizations. For the first time since the period of decolonization in the 1950s, African countries saw the growth of a civil society. Civic initiatives, however, were seriously constrained both legally and politically. Freedom of speech and freedom of association had been extensively limited by law in colonial days. Although the nationalist leaders of the independence struggle had violated these laws to convey their political message, once they came to power, they left the oppressive laws in the books. They further legislated that civic organizations must accept the lead role of a single ruling party. When this type of political system collapsed in the Soviet Union and Eastern Europe, its legitimacy was also quickly questioned in Africa. One of the more bizarre and memorable moments of democratic transition in Africa is what happened in Tanzania. In 1991 the government appointed a special commission to gauge the moods of Tanzanians on this issue. Asked to indicate a preference for continuation of the one-party system or shift to a multi-party system, a majority chose the first option. They wanted continuity and expressed fear that competitive politics would destabilize the country. This majority opinion, however, was unilaterally overruled by the "Father of the Nation," Julius Nyerere, the first president of the country and responsible for the introduction of the one-party system in the first place. He convinced the political leadership that given what was happening to communism it left Tanzania with no choice but to abandon its one-party monopoly of politics and, like countries elsewhere in the world, move on to a multi-party form of politics. Thus, Tanzania embarked on the transition to democracy not because of public pressures or demands but the fiat of a single influential leader challenging public opinion.

In Tanzania and elsewhere in Africa where such legal and political constraints existed, civil society organization gradually took root. With these new actors entering the public scene, donors could not avoid responding to their preferences and needs. Foreign aid no longer remained a prerogative of governments. Non-state actors in the 1990s became beneficiaries of increasing aid flows. External support gave them a sense of status and influence although governments often accused them of being agents of foreign powers. The issue of their position and role in society was politicized and led to discussion of how their role as the voice of the people could be salvaged without appearing a threat to those in power. It was in this context that the concept of an "enabling environment" emerged as a viable challenge to mainstream thinking.

The real challenge in Africa is still to find ways of channeling values from society to the official institutions of the state. To the extent that ideas flow upwards they follow the channels of a clientelist form of politics. Local leaders vie for national positions based on their promise to speak on behalf of their community. Afro-barometer data show that a majority of respondents prefer relying on these informal arrangements rather than official bureaucratic channels (Bratton and Housseou 2014). This practice, however, is discretionary and does not encourage the growth of collective action that involves local households. That is why the concept of an enabling environment makes sense. To incorporate the local dynamics that we have shown in this volume, a policy aimed at establishing an enabling political environment would be needed in which collective action at local level could be spontaneously scaled up to have influence at higher echelons in the policy system (Thindwa 2001).

Africa – the South of the South – quo vadis?

Because of the lasting influences of Natural Society, Africa is a continent that is not only different from the other regions of the Global South but also calls for a different approach to its development challenges. Using such accords as those that have produced the Millennium Development Goals (MDGs) and SDGs, the international community is pushing African countries to adopt strategies and policies that run contrary to how local consumption communities are best developed and eventually integrated into a process that may produce an agrarian transformation. Sub-Saharan Africa is a subregion of the world that poses special challenges.

The paradox that this volume highlights is that while African peasants are very difficult for state and market to induce or "capture" for official development policy purposes, they are also the ones who in a world of climate change may have the most valuable insights and experiences to share. They combine social inclusivity with environmental care and strive for a production level aligned with subsistence needs in ways that an increasing number of people elsewhere in the world are interested in and would like to practice on their own. By sitting on the legacies of Natural Society Africa is at one and the same time both the most underdeveloped part of the world and the repository of ideas and solutions that people in many other countries, including developed ones, view as part of a more sustainable world.

Such a revaluation of African peasants as key actors of "endogenous development" is also in line with the recent paradigm shift of the United Nations' agricultural policy from promoting large-scale commercial agriculture to supporting small-scale family farms, which provide the bulk of world food supply. FAO (2014) estimates that more than 90 percent of all farms in the world are family-owned and occupy 70–80 percent of farmland and produce more than 80 percent of food (in terms of value) the world over. Small-scale family farms are attracting increasing attention not only as viable units of effective and sustainable food production, but also for their potential for conserving natural resources, enhancing biodiversity and, above all, alleviating rural poverty. That is why the UN launched the decade of family farming in 2019 (FAO and IFAD 2019). Behind this bold move away from conventional policies lies a growing threat to local food security and sovereignty posed by multinational companies which have continuously enclosed farmlands and seeds for their own profit in the rural areas of developing countries. Recently, active campaigns have been made by agri-business communities to promote the introduction of either biotech (genetically modified seeds in particular) or ICT ("smart agriculture"), but these state-of-the-art technologies brought from the outside are unlikely to enhance local practices and local ownership of African peasants at the grassroots.

A final conclusion from this volume, therefore, is that rather than the remnants of Natural Society in Africa being destroyed, they must be allowed to continue flourishing on terms that produce local impetus for an agrarian transformation rather than one that is dictated from the outside. Africa is by no means a "hopeless" case. As this volume has demonstrated, it offers its own insights into how humankind has evolved and what in retrospect now seems to be missing.

About the Authors

Yuichiro Fujioka is Lecturer, Faculty of Social and Cultural Studies, Kyushu University, Japan. He has been involved in several research projects focusing on the impact of climate changes on human livelihood, socioeconomic systems and local environment in Namibia and Siberia. One of his recent research interests is changing agro-pastoralism in semi-arid Africa. His publications include "Changes in Natural Resource Use among Owambo Agro-Pastoralists of North-central Namibia Resulting from the Enclosure of Local Frontiers", *African Study Monographs*, Supplementary Issue 40 (2010).

Goran Hyden is Distinguished Professor Emeritus in Political Science at the University of Florida. He has authored many books and articles on African politics and development. These include *Beyond Ujamaa in Tanzania: Underdevelopment and an Uncaptured Peasantry* (University of California Press, 1980), *No Shortcuts to Progress* (Heinemann, 1983), "Governance and Poverty Reduction in Africa", *Proceedings of the National Academy of Sciences*, vol. 104, no. 43 (October 2007), pp. 16751–6, and *African Politics in Comparative Perspective* (Cambridge University Press, 2006 [2nd rev. ed., 2013]).

Kazunobu Ikeya is Professor, National Museum of Ethnology, Japan. He has been coordinating a number of anthropological research projects on hunters-and-gatherers and pastoralist societies in southern Africa, tropical Asia, and the circumpolar region. His main research topics include the sedentarization process among nomadic peoples and ethnic relationships between hunter-gatherers and neighboring cultivators. His publications on this theme include: (co-edited with Robert K. Hitchcock) *Hunter-Gatherers and their Neighbors in Asia, Africa, and South America* (*Senri Ethnological Studies*, no. 94, 2016) and (edited) *Sedentarization among Nomadic Peoples in Asia and Africa* (*Senri Ethnological Studies*, no. 95, 2017).

Yudai Ishimoto is Researcher, Regional Collaboration Center, Aomori Public University, Japan. One of his specific research themes is food security of rural households in semi-arid Africa. His publications on this theme include *A Livelihood System in Sahel: Adaptation and Coping Behavior for Drought and Insect Damage* (Shokado Book Sellers, 2012, in Japanese) and "Food Consumption System as a Survival Strategy in the Semiarid Area: An Empirical Analysis on Agro-pastoralists in the Sahel", *Journal of Arid Land Studies,* vol. 20, no. 2 (2010) (in Japanese).

Noriko Ito is a Researcher at the Policy Research Institute, Ministry of Agriculture, Forestry and Fisheries, Japan. She has been studying agricultural policy and rural development in Africa and Southeast Asia. Her publications include "Market Economic Behavior and Moral Economic Behavior in Farmers' Livelihoods: The Case Study of the Settlers and Their Sons in the Irrigation Scheme Area in Kenya", *Journal of African Studies,* no. 90 (2016) (in Japanese) and "Post-Green Revolution Diffusion Process of Low Input Agricultural Technology in Java, Indonesia: Social Network Analysis of Organic SRI (System of Rice Intensification) Diffusion", *Journal of Agricultural Policy Research,* no. 29 (2018) (in Japanese).

Kenta Sakanashi is Lecturer, Faculty of Agriculture, Ryukoku University, Japan. He has been working on small-scale cacao producers' livelihood and their utilization of natural resources in tropical Africa. He is the author of *Tropical Agriculture and Environmental Conservation in Africa* (Showado Publisher, 2014) (in Japanese) and "Land Use Patterns for Cacao Agroforestry in Southern Cameroon", *African Study Monographs,* vol. 32, no. 4 (2011). One of his recent research projects focuses on African migrants working in rural areas of Japan.

Kazuhiko Sugimura is Professor, Center for Arts and Sciences, Fukui Prefectural University, Japan. He has been organizing several research projects on comparative studies of peasant societies in Africa and other regions. He co-edited several books including (with I. N. Kimambo, G. Hyden, and S. Maghimbi) *Contemporary Perspectives on African Moral Economy* (Dar es Salaam University Press [DUP], 2008), (with S. Maghimbi and I. N. Kimambo) *Comparative Perspectives on Moral Economy: Africa and Southeast Asia* (DUP, 2011), and (with S. Maghimbi and D. Mwamfupe) *Endogenous Development, Moral Economy and Globalization in Agro-pastoral Communities in Central Tanzania* (DUP, 2016).

Yuko Sugiyama is Professor, Faculty of Humanities and Social Sciences, Hirosaki University, Japan. She has been conducting sociocultural studies on local systems of livelihood and their changing process in both rural Africa and rural Japan. One of her specific research themes is how villagers articulate the communal exchange system (based on face-to-face interactions) to the impersonal cash economy. Her publications include the co-authored *Groups: The Evolution of Human Sociality* (Kyoto University Press, 2013).

Tadasu Tsuruta is Professor, Faculty of Agriculture, Kindai University, Japan. He has focused much of his research on the comparative study of peasant societies in Africa and Southeast Asia, one recent research interest being the changing agro-pastoralism in semi-arid Africa. He contributed a chapter on this theme to Jon Abbink (ed.), *The Ecological Crunch in Africa: Growth Narratives vs. Local Environmental Realities* (Palgrave Macmillan, 2018). He also co-authored with Rie Makita, *Fair Trade and Organic Initiatives in Asian Agriculture: The Hidden Realities* (Routledge, 2017).

Haruna Yatsuka is Assistant Professor, Department of International Cooperation and Multicultural Studies, Tsuda University, Japan. She has been working on rural livelihoods and natural resource use in hunter-gatherer communities in Tanzania. Her publications on this theme include "Historical Interaction with Neighbors from the View of Livelihood Change: A Study of the Sandawe of Tanzania", *Senri Ethnological Studies*, no 94 (2016), and "Reconsidering the 'Indigenous Peoples' in the African Context from the Perspective of Current Livelihood and its Historical Changes: the Case of the Sandawe and the Hadza in Tanzania", *African Study Monographs*, vol. 36, no. 1 (2015).

References

Acemoglu, Daron and James A. Robinson 2006. *Economic Origins of Dictatorship and Democracy*. New York: Cambridge University Press.

Ahmed, Abdel Ghaffar M., Alemayehu Azeze, Mustafa Babiker and Diress Tsegaye 2002. *Post Drought Recovery Strategies among the Pastoral Households in the Horn of Africa: A Review*. Addis Ababa: Organization for Social Science Research in Eastern and Southern Africa (OSSREA).

Alexandre, Pierre and Jacquet Binet 1958. *Le groupe dit Pahouin (Fang-Boulou-Beti)*. Paris: Presses Universitaires de France.

Almond, Gabriel A. and James S. Coleman 1960. T*he Politics of Developing Areas*. Princeton, NJ: Princeton University Press.

Amin, Samir 1976. *Unequal Development: An Essay on the Social Formations of Peripheral Capitalism*. New York: Monthly Review Press.

Anderson, James N. 1964. "Land and Society in a Pangasinan Community", in Socorro C. Espiritu and Chester L. Hunt (eds.), *Social Foundations of Community Development: Readings on the Philippines*, pp. 171–99. Manila: R. M. Garcia Publication.

Andersson-Djurfeldt, Agnes 2015. "Multi-Local Livelihoods and Food Security in Africa", *Journal of International Development*, vol. 27, no. 4, pp. 528–45.

Andrews, Matt 2013. *The Limits of Institutional Reform in Development: Changing Rules for Realistic Solutions*. Cambridge: Cambridge University Press.

Ankei, Yuchi 1981. "Agriculture and Economic Activities among the Songola: Swidden Cultivation in Tropical Rainforest, Central Africa", *Kikan Jinruigaku*, vol. 12, no. 1, pp. 96–183 (in Japanese).

Bahuchet, Serge and Henri Guillaume 1982. "Aka-farmer Relations in the Northwest Congo Basin", in Eleanor Leacock and Richard Lee (eds.), *Politics and History in Band Societies*, pp. 189–211. Cambridge: Cambridge University Pres.

Barber, Karin 1991. *I Could Speak until Tomorrow: Oriki, Women and the Past in a Yoruba Town*. Washington, DC: Smithsonian Institution.

Barnard, Alan 2002. "The Foraging Mode of Thought", *Senri Ethnological Studies*, no. 60, pp. 5–24.

Bennett, Albert L. 1898. "Ethnographical Notes on the Fang", *Journal of the Anthropological Institute of Great Britain and Ireland*, vol. 29, no. 2, pp. 66–98.

Bennholdt-Thomsen, Veronica Mies and Maria Mies 2000. *The Subsistence Perspective: Beyond the Globalized Economy*. London: Zed Books.

Berg, Elliott 1993. *Rethinking Technical Cooperation: Reforms for Capacity-Building in Africa*. New York: United Nations Development Programme.

Berk, Gerald and Dennis Galvan 2009. "How People Experience and Change Institutions: A Field Guide to Creative Syncretism", *Theory and Society*, vol. 38, pp. 543–80.

Bernstein, Henry 1979. "African Peasantries: A Theoretical Framework", *Journal of Peasant Studies*, vol. 6, no. 4, pp. 421–43.

Bernstein, Henry and Terence Byres 2001. "From Peasant Studies to Agrarian Change", *Agrarian Change*, vol. 1, no. 1, pp. 1–56.

Bernus, Edmond 1980. "Famines et Sécheresses chez les Touaregs Sahéliens: Les nourritures de substitution", *Africa*, vol. 50, no. 1, pp. 1–7.

Berry, Sara 1993. *No Condition is Permanent: The Social Dynamics of Agrarian Change in sub-Saharan Africa*. Madison, WI: University of Wisconsin Press.

Bird-David, Nurit 1992. "Beyond 'The Hunting and Gathering Mode of Subsistence': Culture-Sensitive Observations on the Nayaka and Other Modern Hunter-Gatherers", *Man (New Series)*, vol. 27, no. 1, pp. 19–44.

Blurton-Jones, Nicholas G. (2016) "Why Do So Few Hadza Farm", in Brian F. Codding and Karen L. Kramer (eds.), *Why Forage? Hunters and Gatherers in the Twenty-First Century*, pp. 113–36. Santa Fe, NM: School for Advanced Research Press.

Booth, David 2015. "Still Watering White Elephants? The Blueprint versus Process Debate Thirty Years On", in A. Mette Kjaer, Lars Engberg-Pedersen and Lars Buur (eds.), *Perspectives on Politics, Production and Public Administration in Africa: Essays in Honor of Ole Therkildsen*, pp. 11–26. Copenhagen: Danish Institute of International Studies.

Booth, David and Frederick Gooloba-Mutebi 2012. "Developmental Patrimonialism: The Case of Rwanda", *African Affairs*, vol. 111, no. 444, pp. 379–403.

Boserup, Ester 1970. *Women's Role in Economic Development*. London: Allen and Unwin.

Boulding, Kenneth 1953. *The Organizational Revolution*. New York: Harper and Row.

Braidwood, Robert J. 1960. "The Agricultural Revolution", *Scientific American*, September 1960, pp. 131–48.

Bratton, Michael and Richard Houessou 2014. "Demand for Democracy is Rising in Africa, but Most Political Leaders Fail to Deliver", *Afro-barometer Policy Paper No. 11*. Johannesburg: IDASA.

Braudel, Fernand 1993. *A History of Civilizations* (translated by Richard Mayne). New York: Penguin Books.

Caldwell, Jack C. 1969. *African Rural-Urban Migration: The Movement to Ghana's Towns*. New York: Columbia University Press.

Carruthers, Ian and Adam Weir 1976. "Rural Water Supplies and Irrigation Development", in Judith Heyer, Joseph K. Maitha, and William M. Senga (eds.), *Agricultural Development in Kenya: An Economic Assessment*, pp. 288–312. Nairobi: Oxford University Press.

Chambers, Robert 1983. *Rural Development: Putting the Last First*. London: Longman.

Chambers, Robert and Gordon Conway 1992. "Sustainable Rural Livelihoods: Practical Concepts for the 21st Century", *Discussion Paper No. 296*. Falmer, East Sussex: Institute of Development Studies.

Chambers, Robert and Jon Moris 1973. *Mwea: An Irrigated Rice Settlement in Kenya*. München: Weltforum Verlag.

Chayanov, Alexander V. 1966. *The Theory of Peasant Economy* (ed. and trans. by Daniel Thorner, Basile Kerblay and R. E. F. Smith). Homewood, IL: Richard D. Irwin.

Childe, Vere Gordon 1936. *Man Makes Himself.* London: Watts and Co.

Cliffe, Lionel 1987. "The Debate on African Peasantries", *Development and Change,* vol. 18, no. 4, pp. 625–35.

Cliffe, Lionel and John S. Saul (eds.) 1973. *Socialism in Tanzania,* vol. I. Nairobi: East African Publishing House.

Colson, Elizabeth 1988. *For Prayer and Profit: The Ritual Economic and Social Importance of Beer in Gwembe District, Zambia 1950–1982.* Stanford, CA: Stanford University Press.

Combe, Jean 1982. "Agroforestry Techniques in Tropical Countries: Potential and Limitations", *Agroforestry System,* vol. 1, no. 1, pp. 13–27.

Connah, Graham 1987. *African Civilizations.* Cambridge: Cambridge University Press.

Connelly, W. Thomas and Miriam S. Chaiken 2000. "Intensive Farming, Agro-Diversity, and Food Security under Conditions of Extreme Population Pressure in Western Kenya", *Human Ecology,* vol. 28, no. 1, pp. 19–51.

Coquery-Vidrovitch, Catherine 1978. "Research on an African Mode of Production", in D. Seddon (ed.) *Relations of Production: Marxist Approaches to Economic Anthropology,* pp. 261–88. London: Frank Cass.

Coquery-Vidrovitch, Catherine 1988. *Africa: Endurance and Change South of the Sahara* (trans. by D. Maisel). Berkeley, CA: University of California Press.

Coulson, Andrew 1982. *Tanzania: A Political Economy.* Oxford: Oxford University Press.

Dalton, George 1972. "Peasantries in Anthropology and History", *Current Anthropology,* vol. 13, nos 3–4, pp. 411–15.

den Biggelaar, Christophel 1991. "Farming Systems Development: Synthesizing Indigenous and Scientific Knowledge Systems", *Agriculture and Human Values,* vol. 8, nos 1–2, pp. 25–36.

Disch, Arne, Endre Vigeland and Geir Sundet 2009. "Anti-Corruption Approaches: A Literature Review", *Sida Evaluation Report No 2009:1.* Stockholm: Swedish International Development Authority.

Djurfeldt, Göran 2005. "Global Perspectives on Agricultural Development", in Göran Djurfeldt, Hans Holmén, Magnus Jirström and Rolf Larsson (eds.), *The African Food Crisis: Lessons from the Asian Green Revolution,* pp. 9–24. Wallingford, UK: CABI Publishing.

Dounias, Edmond 1993. "Perception and Use of Wild Yams by the Baka Hunter-gatherers in South Cameroon", in Claude M. Hladik, Annette Hladik, Olga F. Linares, Helen Pagezy, Alison Semple and Malcolm Hadley (eds.), *Tropical Forests, People and Food: Biocultural Interactions and Applications to Development,* pp. 621–32. Paris: UNESCO.

Easton, David 1965. *A Systems Analysis of Political Life.* New York: Wiley.

Ekeh, Peter 1975. "Colonialism and the Two Publics in Africa: A Theoretical Statement", *Comparative Studies in Society and History,* vol. 17, no. 1, pp. 91–112.

Elson, Robert. E. 1997. *The End of the Peasantry in Southeast Asia: A Social and Economic History of Peasant Livelihood, 1800–1990s.* London: Macmillan.

Ericksen, Annika 2010. "Productions of Space in Tuareg History: Power, Marginalization, and Resistance", *Arizona Anthropologist,* vol. 20, pp. 60–80.

Evers, Hans Dieter (ed.) 1969. *Loosely Structured Social Systems: Thailand in Comparative Perspective.* New Haven, CT: Yale University Press.

FAO (Food and Agriculture Organization) 2014. *The State of Food and Agriculture: Innovation in Family Farming.* Rome: FAO.

FAO (Food and Agriculture Organization) and IFAD (International Fund for Agricultural Development) 2019. *United Nations Decade of Family Farming 2019–2028. Global Action Plan.* Rome: FAO.

Fujioka, Yuichiro 2010. "Changes in Natural Resource Use among Owambo Agro-Pastoralists of North-Central Namibia Resulting from the Enclosure of Local Frontiers", *African Study Monographs,* Supp. Issue 40, pp. 129–54.

Fujioka, Yuichiro 2014. "Rural Entrepreneurs and Social Connections: The Management of Cattle Posts and Interactions among Farmers in North-Central Namibia", *MILA Special Issue on Exploring African Potentials: The Dynamics of Action, Living Strategy and Social Order in Southern Africa,* vol. 12, pp. 25–38.

Fujita, Koichi 1990. "A Study of Labor Custom in Rural Java: Case Studies in Rural Non-irrigated Western Java", *Quarterly Journal of Agricultural Economy,* vol. 44, no. 3, pp. 1–53 (in Japanese).

Fukui, Katsuyoshi 1969a. "Ecological Observation of the Agro-Pastoral People: On the Migration and Settlement of the Iraqw in Tanzania", *Journal of African Studies,* no. 19, pp. 1–18 (in Japanese).

Fukui, Katsuyoshi 1969b. "The Subsistence Economy of the Agrico-pastoral Iraqw", *Kyoto University African Studies,* vol. 4, pp. 41–76.

Gallais, Jean 1977. *Stratégies pastorales et agricoles des Sahéliens durant la sécheresse 1969–1974; elevage et contacts entre pasteurs et agriculteurs.* Talence: Centre d'Études de Géographie Tropicale.

Geertz, Clifford 1963. *Agricultural Involution: The Process of Ecological Change in Indonesia.* Berkeley, CA: University of California Press.

Goldman, Michael 2005. *Imperial Nature: The World Bank and Struggles for Social Justice in the Age of Globalization.* New Haven, CT: Yale University Press.

Goody, Jack 1971. *Technology, Tradition and the State in Africa.* London: Oxford University Press.

Goody, Jack 1976. *Production and Reproduction: A Comparative Study of the Domestic Domain.* Cambridge: Cambridge University Press.

Government Spending Watch 2015. *Financing the Sustainable Development Goals: Lessons from Government Spending on the MDGs.* London: Development Finance International and Oxfam.

Guyer, Jane I. 1993. "Wealth in People and Self-Realization in Equatorial Africa", *Man (New Series),* vol. 28, no. 2, pp. 243–65.

Habermas, Jurgen 1987. *Theory of Communicative Action,* Vol. II. *Lifeworld and System: A Critique of Functionalist Reason* (trans. by Thomas McCarthy). Boston, MA: Beacon Press.

Hallpike, Christopher Robert 2011. *On Primitive Society: And Other Forbidden Topics.* Bloomington, IN: AuthorHouse Publishing.

Hart, Keith 1982. *The Political Economy of West African Agriculture.* Cambridge: Cambridge University Press.

Hayami, Yujiro and Masao Kikuchi 1981. *Asian Village Economy at the Crossroad: An Economic Approach to Institutional Change.* Tokyo: University of Tokyo Press.

Hayashi, Koji 2000. "Hunting and Gathering Activities of the Baka in South-Eastern Cameroon: The Actual Condition and Contemporary Significance", *Human and Science,* no. 14, pp. 27–38 (in Japanese).

Henley, David 2012, "The Agrarian Roots of Industrial Growth: Rural Development in South-East Asia and sub-Saharan Africa", *Development Policy Review,* vol. 30, no. 51, pp. 525–47.

Herbst, Jeffrey 2000. *States and Power in Africa: Comparative Lessons in Authority and Control.* Princeton, NJ: Princeton University Press.

Herbst, Jeffrey 2014. *States and Power in Africa: Comparative Lessons in Authority and Control* (new ed.). Princeton, NJ: Princeton University Press.

Hill, Polly 1963. *Migrant Cocoa Farmers of Southern Ghana.* London: Cambridge University Press.

Hirano, Katsumi 2013. *The Economy in the African Continent.* Tokyo: Chuokoron-sha (in Japanese).

Hirose, Shohei and Toshiyuki Wakatsuki (eds.) 1997. *Restoration of Ecological Environment and Villages in West African Savanna.* Tokyo: Norin Tokei Kyokai (in Japanese).

Hirschman, Albert O. 1963. *Journeys Toward Progress: Studies of Economic Policymaking in Latin America.* New York: 20th Century Fund.

Hirschman, Albert O. 1970. *Exit, Voice and Loyalty: Responses to Decline in Firms, Organizations and States.* Cambridge, MA: Harvard University Press.

Hodge, Joseph Morgan 2007. *Triumph of the Expert: Agrarian Doctrines of Development and the Legacies of British Colonialism.* Athens, OH: Ohio University Press.

Huizer, Gerrit 1973. *Peasant Rebellion in Latin America: The Origins, Forms of Expression, and Potential in Latin American Peasant Unrest.* Harmondsworth: Penguin.

Huntington, Samuel P. 1965. "Political Development and Political Decay", *World Politics,* vol. 17, no. 3, pp. 386–430.

Hyden, Goran 1980. *Beyond Ujamaa in Tanzania: Underdevelopment and an Uncaptured Peasantry.* Berkeley, CA: University of California Press.

Hyden, Goran 1983. *No Shortcuts to Progress: Development Management in Perspective.* London: Heinemann Educational Books.

Ichikawa, Mitsuo 1986. "Ecological Bases of Symbiosis, Territoriality and Intraband Cooperation among the Mbuti Pygmies", *Sprache und Geschichte in Afrika,* vol. 7, no. 1, pp. 161–88.

Ichikawa, Mitsuo 2005. "Food Sharing and Ownership among the Central African Hunter-Gatherers: An Evolutionary Perspective", in Thomas Widlok and Wolde Tadesse (eds.) *Property and Equality,* pp. 151–64. Oxford: Berghahn Books.

Ikeya, Kazunobu 1993. "Goat Raising among the San in the Central Kalahari", *African Study Monographs,* vol. 14, no. 1, pp. 39–52.

Ikeya, Kazunobu 1996. "Dry Farming among the San in the Central Kalahari." *African Study Monographs,* Supp. Issue 22, pp. 85–100.

Ikeya, Kazunobu 1999. "Historical Dynamics of the Socioeconomic Relationships between the Nomadic San and Rural Kgalagadi", *Botswana Notes and Records,* no. 31, pp. 19–33.

Ikeya, Kazunobu 2001. "Some Changes among the San under the Influence of Relocation Plan in Botswana", in David G. Anderson and Kazunobu Ikeya (eds.), *Parks, Property, and Power: Managing Hunting Practice and Identity within State Policy Regimes* (*Senri Ethnological Studies,* no. 59), pp. 183–98. Osaka: National Museum of Ethnology.

Ikeya, Kazunobu 2002. *Hunter-Gatherers in the State Society: An Historical Ethnography of the San in the Kalahari.* Osaka: National Museum of Ethnology (in Japanese).

Ikeya, Kazunobu 2007. "Subsistence and Sharing among the Kalahari Hunter-Gatherers", *Journal of African Studies,* no. 70, pp. 91–101 (in Japanese).

Ikeya, Kazunobu 2018. "Settlement Patterns and Sedentarization among the San in the Central Kalahari (1930–1996)", in R. Fleming Puckett and Kazunobu

Ikeya (eds.), *Research and Activism among the Kalahari San Today: Ideals, Challenges, and Debates* (*Senri Ethnological Studies*, no. 99), pp. 177–96. Osaka: National Museum of Ethnology.

Iliffe, John 1995. *Africans: The History of a Continent.* Cambridge: Cambridge University Press.

Imanishi, Kinji 2002. *A Japanese View of Nature: The World of Living Things* (compiled and ed. by Pamela J. Asquith). London: Routledge.

Ishii, Yoko 2007. *Ethnography of Development Frontier: People Living in Irrigation Scheme Area in East Africa.* Tokyo: Ochanomizu-shobo (in Japanese).

Itani, Jun'ichiro 1988. "The Origin of Human Equality", in Michael R. A. Chance (ed.), *Social Fabrics of the Mind*, pp. 137–56. Mahwah, NJ: Lawrence Erlbaum Associates.

Ito, Noriko 2016. "Market Economic Behavior and Moral Economic Behavior in Farmers' Livelihoods: Case Study of the Settlers and Their Sons in the Irrigation Scheme Area in Kenya", *Journal of African Studies*, vol. 90, pp. 15–28 (in Japanese).

Ito, Noriko 2017a. "Mechanism and Function of Food Sharing Practices in Rural Africa: Case Study of Rice Consumption Process of Farmers in the Irrigation Scheme Area in Kenya", *Journal of Agricultural Policy Research*, vol. 27, pp. 1–24 (in Japanese).

Ito, Noriko 2017b. "Rice Transaction among Farmers in Kenya: A Study on Social Background of Rice Selling in Mwea Irrigation Scheme Area", *Journal of Food System Research*, vol. 24, no. 3, pp. 315–20 (in Japanese).

JICA (Japan International Cooperation Agency) 2015. *Joint Mid-Term Review Report on the Japanese Technical Cooperation for Race-Based and Market-Oriented Agriculture Promotion Project in the Republic of Kenya.* Tokyo: JICA (in Japanese).

JICA (Japan International Cooperation Agency) 2018. *SHEP (Smallholder Horticulture Empowerment Project) Handbook for Extension Staff: A Practical Guide to the Implementation of SHEP Approach.* Tokyo: JICA.

Joseph, Richard A. 1977. *Radical Nationalism in Cameroon: Social Origins of the U.P.C. Rebellion.* Oxford: Clarendon Press.

Kabutha, Charity and Clifford Mutero 2002. "From Government to Farmer-Managed Smallholder Rice Schemes: The Unresolved Case of the Mwea Irrigation Scheme in Kenya", in Herbert G. Blank, Clifford M. Mutero and Hammond Murray-Rust (eds.), *The Changing Face of Irrigation in Kenya: Opportunities for Anticipating Changes in Eastern and Southern Africa*, pp. 127–37. Colombo, Sri Lanka: International Water Management Institute.

Kakeya, Makoto 1976. "Subsistence Ecology of the Tongwe, Tanzania", *Kyoto University African Studies*, vol. 10, pp. 143–212.

Kakeya, Makoto 1982. "Curing Ritual of the Tongwe Traditional Doctor: Its Process and Logic", *African Study Monographs*, Supp. Issue 1, pp. 105–39.

Kakeya, Makoto 1986. "Livelihood Structure of Traditional Peasants: A Case of the Tongwe", in Juni'chiro Itani and Jiro Tanaka (eds.), *Anthropology of Natural Society*, pp. 218–42. Kyoto: Academia Publisher (in Japanese).

Kakeya, Makoto 2018. *Collected Works of Makoto Kakeya*, vol. III. Kyoto: Kyoto University Press (in Japanese).

Kakeya, Makoto and Yuko Sugiyama 1985. "*Citemene*, Finger Millet and Bemba Culture", *African Study Monographs*, Supp. Issue 4, pp. 1–24.

Kakeya, Makoto and Yuko Sugiyama 1987. "Agricultural Change and Its Mechanism in the Bemba Villages of Northeastern Zambia", *African Study Monographs*, Supp. Issue 6, pp. 1–13.

Kakeya, Makoto, Yuko Sugiyama and Shuichi Oyama 2006. "The *Citemene* System, Social Leveling Mechanism, and Agrarian Changes in the Bemba Villages of Northern Zambia: An Overview of 23 years of 'Fixed-Point' Research", *African Study Monographs*, vol. 27, no. 1, pp. 27–38.

Kano, Hiroyoshi 1979. *Pagelaran: Wealth and Poverty in the Villages in Eastern Java.* Tokyo: Institute of Development Economics (in Japanese).

Kano, Hiroyoshi 1981. *Sawahan: Rural Areas in Central Java under the Developmental Regime.* Tokyo: Institute of Development Economics (in Japanese).

Karatani, Kojin 2014. *The Structure of World History: From Modes of Production to Modes of Exchange.* Durham, NC and London: Duke University Press.

Kasfir, Nelson 1986. "Are African Peasants Self-sufficient?", *Development and Change*, vol. 17, no. 2, pp. 335–57.

Kenyatta, Jomo 1978. *Facing Mount Kenya: The Tribal Life of the Gikuyu.* Nairobi: Heinemann Kenya.

Khan, Musthaq 2017. "Anti-Corruption in Bangladesh: A Political Settlements Analysis", *Anti-Corruption Evidence Research Consortium Working Paper No. 3.* London: School of Oriental and African Studies, University of London.

Kiggundu, Moses 1991. "The Challenges of Management Development in sub-Saharan Africa", *Journal of Management Development*, vol. 10, no. 6, pp. 42–57.

Kimambo, Isaria N., Goran Hyden, Samuel Maghimbi and Kazuhirko Sugimura (eds.) 2008. *Contemporary Perspectives on African Moral Economy.* Dar es Salaam: Dar es Salaam University Press.

Kimura, Daiji 1992. "Daily Activities and Social Association of the Bongando in Central Zaire", *African Study Monographs*, vol. 13, no. 1, pp. 1–33.

Kitanishi, Koichi 2000. "The Aka and Baka: Food Sharing among Two Central Africa Hunter-Gatherer Groups", in George W. Wenzel, Grete Hovelsrud-Broda and Nobuhiro Kishigami (eds.), *The Social Economy of Sharing: Resource Allocation and Modern Hunter-Gatherers, Senri Ethnological Studies*, no. 53, pp. 149–69. Osaka: National Museum of Ethnology.

Kitanishi, Koichi 2003. "Cultivation by the Baka Hunter-Gatherers in the Tropical Rain Forest of Central Africa", *African Study Monographs*, Supp. Issue 28, pp. 143–57.

Koike, Satoshi and Teitaro Kitamura 1987. "Agricultural Work and Social Relations: A Case from R-village in East Java", *Journal of Rural Planning Association*, vol. 6, no. 2, pp. 26–35 (in Japanese).

Koponen, Johannes 1988. *People and Production in Late Pre-colonial Tanzania: History and Structures.* Uppsala: Scandinavian Institute of African Studies.

Kopytoff, Igor (ed.) 1987. *The African Frontier: The Reproduction of Traditional African Societies.* Bloomington, IN: Indiana University Press.

Leonard, David K. 1977. *Reaching the Peasant Farmer: Organisation Theory and Practice in Kenya.* Chicago, IL: University of Chicago Press.

Levy, Brian 2013. *Working with the Grain: Integrating Governance and Growth in Development Strategies.* New York: Oxford University Press.

Lindberg, Staffan 2003. "It's Our Time to Chop: Do Elections in Africa Feed Neo-Patrimonialism rather than Counteract It?", *Democratization*, vol. 10, no. 2, pp. 111–40.

Lindblom, Charles E. 1959. "The Science of 'Muddling Through,'" *Public Administration Review*, vol. 19, no. 2, pp. 79–88.

Lindblom, Charles E. 1990. *Inquiry and Change: The Troubled Attempt to Understand and Change Society.* New Haven, CT: Yale University Press.

Loiske, Vesa-Matti 2004. "Institutionalized Exchange as a Driving Force in Intensive Agriculture", in Mats Widgren and John E. G. Sutton (eds.), *Islands of Intensive Agriculture in Eastern Africa*, pp. 105–13. Oxford: James Currey.

Luxemburg, Rosa 2003. *The Accumulation of Capital* (trans. by Agnes Schwarzschild). New York: Routledge.

McIntire, John, Daniel Bourzat, and Prabhu Pingali 1992. *Crop-livestock Interaction in sub-Saharan Africa*. Washington, DC: The World Bank.

Maghimbi, Samuel, Kazuhiko Sugimura and Davis G. Mwamfupe (eds.) 2016. *Endogenous Development, Moral Economy and Globalization in Agropastoral Communities in Central Tanzania*. Dar es Salaam: Dar es Salaam University Press.

Maruyama, Junko 2003. "The Impacts of Resettlement on Livelihood and Social Relationships among the Central Kalahari San", *African Study Monographs*, vol. 24, no. 4, pp. 223–45.

Maruyama, Junko. 2010. *The Bushmen in Transition: Between Development Policy and Indigenous People's Movement*. Kyoto: Sekaishiso-sha (in Japanese).

Mati, Bancy, Raphael Wanjogu, Benson Odongo and Patrick G. Home 2011. "Introduction of the System of Rice Intensification in Kenya: Experiences from Mwea Irrigation Scheme", *Paddy and Water Environment*, vol. 9, no. 1, pp. 145–54.

Matsui, Takeshi 2004. "Minor Subsistence and Daily Life", in Ryutaro Otsuka, Takeshi Matsui and Toru Shinohara (eds.), *Toward Renewed Human-Environment Systems: The Impact of Development on Island Habitats* No. 4, pp. 61–84. Tokyo: University of Tokyo Press (in Japanese).

Mazrui, Ali A. 1968. "From Social Darwinism to Current Theories of Modernization", *World Politics*, vol. 21, no. 1, pp. 69–83.

Meillassoux, Claude 1967. "Recherche d'un niveau de determination dans la société cynégétique", *L'Homme et la société*, no. 6, pp. 95–106.

Meillassoux, Claude 1975. *Femmes, greniers, capitaux*. Paris: Maspero.

Mendelsohn, John, Alice Jarvis and Tony Robertson 2013. *A Profile and Atlas of the Cuvelai–Etosha Basin*. Windhoek: Raison and Gondwana Collection.

Miers, Suzanne and Igor Kopytoff 1977. *Slavery in Africa: Historical and Anthropological Perspectives*. Madison, WI: University of Wisconsin Press.

Ministry of Economy and Finances, Republic of Cameroon. 2000. *Cameroon Statistical Yearbook*. Yaoundé: National Institute of Statistics.

Montgomery, John D. 1987. "Probing Managerial Behavior: Image and Reality in Southern Africa", *World Development*, vol. 15, no. 4, pp. 518–34.

Moore, Barrington Jr. 1966. *Social Origins of Dictatorship and Democracy: Lord and Peasant in the Making of the Modern World*. Boston, MA: Beacon Press.

Morris-Suzuki, Tessa 1998. *Re-Inventing Japan: Time, Space, Nation*. New York: M. E. Sharpe.

Mortimore, Michael J. and William M. Adams 2001. "Farmer Adaptation, Change, and 'Crisis' in the Sahel", *Global Environmental Change*, vol. 11, pp. 49–57.

Murdock, George Peter 1959. *Africa: Its Peoples and Their Culture History*. New York, Toronto, ON and London: McGraw-Hill Book Company.

Nakada, Yoshiaki 1995. "Surplus Rice and Seasonal Labor Migration: A Case Study at a Village in Yasothon Province, Northeast Thailand", *Southeast Asian Studies*, vol. 32, no. 4, pp. 523–48 (in Japanese).

Nakada, Yoshiaki 1996. "When Does a Farmer Sell Rice? A Case Study in a Village in Yasothon Province, Northeast Thailand", *Southeast Asian Studies*, vol. 33, no. 4, pp. 609–24.

Nakamura, Hisahi 1976. *Accumulation and Interchange of Labour: An Inquiry into the Non-Market Economy in a South Indian Village.* Tokyo: Institute of Developing Economics.

Nakao, Sasuke 1966. *The Origin of Domesticated Plants and Agriculture.* Tokyo: Iwanami-shoten (in Japanese).

Nakao, Sasuke 1969. *From the Niger to the Nile: A Journey in Search of the Origin of Agriculture.* Tokyo: Kodan-sha (in Japanese).

Ndoye, Ousseynou and David Kaimowitz 2000. "Macro-Economics, Markets and the Humid Forests of Cameroon, 1967–1997", *Journal of Modern African Studies,* vol. 38, no. 2, pp. 225–53.

Newman, James L. 1968. *Geography and Subsistence Change among the Sandawe.* PhD Thesis, University of Minnesota.

Newman, James L. 1970. *The Ecological Basis for Subsistence Change among the Sandawe of Tanzania.* Washington, DC: National Academy of Science.

Newman, James L. 1980. "Dimensions of Sandawe Diet", in John R. K. Robson (ed.), *Food, Ecology and Culture, Reading in the Anthropology of Dietary Practices,* pp. 27–33. New York: Routledge.

Niane, Djibril Tamsir (ed.) 1984. *General History of Africa, IV: Africa from the Twelfth to the Sixteenth Century.* London and Berkeley, CA: Heinemann Educational Books and University of California Press.

Nisbet, Robert 1969. *Social Change and History: Aspects of the Western History of Development.* New York: Oxford University Press.

Njeru, Timothy N, Yukichi Mano and Keijiro Otsuka 2016. "Role of Access to Credit in Rice Production in sub-Saharan Africa: The Case of Mwea Irrigation Scheme in Kenya", *Journal of African Economies,* vol. 25, no. 2, pp. 300–21.

North, Douglass C. 1990. *Institutions, Institutional Change and Economic Performance.* New York: Cambridge University Press.

North, Douglass C., John J. Wallis and Barry R. Weingast 2009. *Violence and Social Orders: A Conceptual Framework for Interpreting Recorded Human History.* New York: Cambridge University Press.

Oishi, Takanori 2012. "Cash Crop Cultivation and Interethnic Relations of the Baka Hunter-Gatherers in Southeastern Cameroon", *African Study Monographs,* Supp. Issue 43, pp. 115–36.

Osaki, Masakazu 1990. "The Influence of Sedentism on Sharing among the Central Kalahari Hunter-Gatherers", *African Study Monographs,* Supp. Issue 12, pp. 59–87.

Ostrom, Elinor 1990. *Governing the Commons: The Evolution of Institutions for Collective Action.* Cambridge: Cambridge University Press.

Ouedraogo, Mathieu and Toubou Ripama 2009. *Etat et structure de la population, analyse des resultats definitifs, recensement general de la population et de l'habitation (RGPH) de 2006.* Ouagadougou: Institut National de la Statistique et de la Démographie.

Oyama, Shuichi 2013. "Miombo Woodland and *Chitemene* Shifting Cultivation of Bemba People in Northern Zambia: Specifying and Calculating a Concept of Optimum Fallow Period", *Ökumene Research,* vol. 2, pp. 21–37 (in Japanese).

Oyama, Shuichi and Tomoki Y. Takamura 2001. "Agrarian Changes and Coping Strategies for Subsistence of Bemba Shifting Cultivators in Northern Zambia in the mid-1990s", *Japanese Journal of Tropical Agriculture,* vol. 45, no. 2, pp. 84–97.

Paige, Jeffrey M. 1975. *Agrarian Revolution: Social Movements and Export Agriculture in the Underdeveloped World.* New York: Free Press.

Parkin, David 1972. *Palms, Wine and Witnesses.* San Francisco, CA: Chandler Publishing House.

Patin, Etienne G., Guillaume Laval, Luis B. Barreiro, Antonio Salas, Ornella Semino, Silvana Santachiara-Benerecetti, Kenneth K. Kidd, Judith R. Kidd, Lolke Van der Veen, Jean-Marie Hombert, Antoine Gessain, Alain Froment, Serge Bahuchet, Evelyne Heyer and Lluís Quintana-Murci 2009. "Inferring the Demographic History of African Farmers and Pygmy Hunter–Gatherers Using a Multilocus Resequencing Data Set", *PLoS Genetics* vol. 5, no. 4, e1000448.

Powell, J. Mark, Felix N. Ikpe, Zana C. Somda and Sergio Fernandez-Rivera 1998. "Urine Effects on Soil Chemical Properties and the Impact of Urine and Dung on Pearl Millet Yield", *Experimental Agriculture*, vol. 34, no. 3, pp. 259–76.

Richards, Audrey I. 1931. Field notes (London School of Economics archives).

Richards, Audrey I. 1939. *Land, Labour and Diet Northern Rhodesia.* London: Oxford University Press.

Richards, Paul 1985. *Indigenous Agricultural Revolution: Ecology and Food Production in West Africa.* Boulder, CO: Westview Press.

Rigby, Peter 1969. *Cattle and Kinship among the Gogo: A semi-pastoral Society of Central Tanzania.* Ithaca, NY: Cornell University Press.

Rodney, Walter 1972. *How Europe Underdeveloped Africa.* London: Bogle-L'Ouverture Publications.

Roussel, Louis 1971. "Employment Problems and Policies in the Ivory Coast", *International Labour Review*, vol. 104, no. 6, pp. 505–25.

Rudolph, Lloyd I. and Susanne H. Rudolph 1967. *The Modernity of Tradition: Political Development in India.* Chicago, IL: University of Chicago Press.

Ruf, François O. 2011. "The Myth of Complex Cocoa Agroforests: The Case of Ghana", *Human Ecology*, vol. 39, no. 3, pp. 373–88.

Rweyemamu, Anthony and Goran Hyden (eds.) 1975. *A Decade of Public Administration in Africa.* Nairobi: East African Literature Bureau.

Sahlins, Marshall 1972. *Stone Age Economics.* New York: Aldine de Gruyter.

Sakanashi, Kenta 2011. "Land Use Patterns for Cacao Agroforestry in Southern Cameroon", *African Study Monographs*, vol. 32, no. 4, pp. 135–55.

Sandbrook, Richard 1982. "The Politics of Basic Needs: Urban Aspects of Assaulting Poverty in Africa", *Canadian Journal of Political Science*, vol. 5, no. 1, pp. 1–26.

Sato, Hiroaki 1983. "Hunting of the Boyela, Slash-and-Burn Agriculturalists in the Central Zaire Forest", *African Study Monographs*, vol. 4, pp. 1–54.

Sato, Shun 1992. "The Camel Trust System in the Rendille Society of Northern Kenya", *African Study Monographs*, vol. 13, no. 2, pp 69–89.

Schumacher, Ernst F. 1973. *Small is Beautiful: Economics as if People Mattered.* New York: Harper & Row.

Schumpeter, Joseph A. 1981[1934]. *The Theory of Economic Development.* London: Routledge.

Scott, James C. 1976. *The Moral Economy of the Peasant: Rebellion and Subsistence in Southeast Asia.* New Haven, CT: Yale University Press.

Scott, James C. 2009. *The Art of Not Being Governed: An Anarchist History of Upland Southeast Asia.* New Haven, CT and London: Yale University Press.

Scott, James C. 2017. *Against the Grain: A Deep History of the Earliest States.* New Haven, CT and London: Yale University Press.

Sekimoto, Teruo 1976. "Rituals of Food Exchange in Rural Central Java", *Bulletin of the National Museum of Ethnology*, vol. 1, no. 3, pp. 457–504 (in Japanese).

Shanin, Theodore 1973. "The Nature and Logic of the Peasant Economy: A Generalisation", *Journal of Peasant Studies*, vol. 1, no. 1, pp. 63–80.

Shigeta, Masayoshi and Morie Kaneko 2017. "ZAIRAICHI (Local Knowledge) as the Manners of Co-Existence: Encounters between the Aari Farmers in South-western Ethiopia and the 'Other' ", in Gebre Yntiso et al. (eds.), *African Virtues in the Pursuit of Conviviality: Exploring Local Solutions in Light of Global Prescriptions*, pp. 311–38. Bamenda: Langaa RPCIG.

Shikata, Kagari 2013. *The Potential of Swidden Agriculture: Agro-Ecology in African Rainforest*. Kyoto: Showado (in Japanese).

Shils, Edward 1963. "On the Comparative Study of the New States", in Clifford Geertz (ed.), *Old Societies and New States*, pp. 1–26. New York: The Free Press.

Shimada, Yoshihito 2006. "Les royaumes Africains comme systémes régionaux: zones écologiques et formations étatiques en Afrique", in Yoshihito Shimada (ed.), *Africa Traditional Kingdoms Studies III*, pp. 43–56. Nagoya: Nagoya University.

Siiskonen, Harri 1990. *Trade and Socioeconomic Change in Ovamboland, 1850–1906*. Helsinki: Societas Historica Fennica.

Simon, Herbert A. 1957. "A Behavioral Model of Rational Choice", in H. A. Simon, *Models of Man, Social and Rational: Mathematical Essays on Rational Human Behavior in a Social Setting*. New York: Wiley.

Soda, Osamu 2006. *Philosophy of Agricultural Science: A Japanese Perspective*. Melbourne, VIC: Transpacific Press.

Sonwa, Denis. J., Bernard A. Nkongmeneck, Stephan F. Weise, Tchatat Maturin, Akin A. Adesina and Marc J. J. Janssens 2007. "Diversity of Plants in Cocoa Agroforests in the Humid Forest Zone of Southern Cameroon", *Biodiversity and Conservation*, vol. 16, no. 8, pp. 2385–427.

Suehara, Tatsuro 1983. "Labour Exchange among the Tembo", *African Study Monographs*, vol. 3, pp. 59–69.

Sugimura, Kazihuko 2004. *The Livelihood of African Peasants: An Inter-regional Comparison on Organizational Principle*. Kyoto: Sekaishiso-sha (in Japanese).

Sugimura, Kazihuko 2006. "The Changing Practices of 'Kibarua' Employment", *African Studies Quarterly*, vol. 9, nos 1 & 2, pp. 67–78.

Sugiyama, Yuko 1987. "Maintaining A Life of Subsistence in the Bemba Village of Northeastern Zambia", *African Study Monographs*, Supp. Issue 6, pp. 15–32.

Sugiyama, Yuko 1988. "The Bemba Women of Northern Zambia: Life Strategies and Subsistence Activity among Slash-and-Burn Cultivators", *Japanese Journal of Ethnology*, vol. 53, no. 1, pp. 31–57 (in Japanese).

Sugiyama, Yuko 1992. "The Development of Maize Cultivation and Changes in the Village Life of the Bemba of Northern Zambia", *Senri Ethnological Studies*, no. 31, pp. 173–201. Osaka: National Museum of Ethnology.

Sugiyama, Yuko 2001. "The Agricultural Policy Changes and the Bemba Villages of Northern Zambia, 1983–2000", in Tsutomu Takane (ed.), *Rural Transformation in Contemporary Africa*, pp. 223–77. Tokyo: Institute of Developing Economics (in Japanese).

Sumberg, James 1998. "Mixed Farming in Africa: The Search for Order, the Search for Sustainability", *Land Use Policy*, vol. 15, no. 4, pp. 293–317.

Sunderlin, William D., Ousseynou Ndoye, Herinette Bikie, Nadine Laporte, Benoit Mertens and Jacques Pokam 2000. "Economic Crisis, Small-Scale Agriculture, and Forest Cover Change in Southern Cameroon", *Environmental Conservation*, vol. 27, no. 3, pp. 284–90.

Takeuchi, Kiyoshi 2014. "Interethnic Relationships between Pygmies and Farmers", in Barry S. Hewlett (ed.) *Hunter-Gatherers of the Congo Basin*, pp. 299–320. New Brunswick, NJ: Transaction Publishers.

Tanaka, Jiro 1980. *The San, Hunter-Gatherers of the Kalahari*. Tokyo: University of Tokyo Press.

Tanaka, Jiro 1982. "Adaptation to Arid Environment: A Comparative Study of Hunter-Gatherers and Pastoralists in Africa", *African Study Monographs*, Supp. Issue 1, pp. 1–12.

Tanaka, Jiro 1986. "Changing Process Followed by Sedentarization and Population Concentration: A Case of Central Bushmen", in Jun'ichiro Itani and Jiro Tanaka (eds.), *Anthropology in Natural Society*, pp. 313–48. Kyoto: Academia Publisher (in Japanese).

Therkildsen, Ole 1988. *Watering White Elephants: Lessons from Donor Funded Planning and Implementation of Rural Water Supplies in Tanzania*. Uppsala: Scandinavian Institute of African Studies.

Thindwa, Jeff 2001. *Enabling Environment for Civil Society in CDD Projects*. Washington, DC: The World Bank.

Thompson, Edward P. 1971. "The Moral Economy of the English Crowd in the 18th Century", *Past & Present*, vol. 50, pp. 76–136.

Tiffen, Mary, Michael Mortimore, and Francis Gichuki 1994. *More People, Less Erosion: Environmental Recovery in Kenya*. Nairobi: ACTS Press.

Tilly, Charles 1975. *The Formation of National States in Western Europe*. Princeton, NJ: Princeton University Press.

Tipps, Dean C. 1973. "Modernization Theory and the Comparative Study of Societies: A Critical Perspective", *Comparative Studies in Society and History*, vol. 15, no. 2, pp. 199–226.

Tishkoff, Sara A., M. K. Gonder, B. M. Henn, H. Mortensen, A. Knight, C. Gignoux, N. Fernandopulle, G. Lema, T. B. Nyambo, U. Ramakrishnan, F. A. Reed and J. L. Mountain. 2007 "The History of Click-Speaking Population of Africa Inferred from mtDNA and Y Chromosome Genetic Variation", *Molecular Biological Evolution*, vol. 24, no. 10, pp. 2180–95.

Tomosugi, Takashi 1975. "Rice Growing and Society in the Chao Phraya Delta", in Yoneo Ishii (ed.), *Thailand: A Rice Growing Society*. Tokyo: Sobun-sha (in Japanese).

Tomosugi, Takashi 1980. *A Structural Analysis of Thai Economic History*. Tokyo: Institute of Developing Economies.

Tomosugi, Takashi 1995. *Changing Features of a Rice-Growing Village in Central Thailand: A Fixed-Point Study from 1967 to 1993*. Tokyo: Center for East Asian Cultural Studies for UNESCO, The Tokyo Bunko.

Tsuruta, Tadasu 2004. "Changing Values in Reciprocity: A Case Study of Commercialized Thai Farmers", *Tanzanian Journal of Population Studies and Development*, vol. 11, no. 2, pp. 103–16.

Tsuruta, Tadasu 2008. "Between Moral Economy and Economy of Affection", in Isaria Kimambo, Goran Hyden, Samuel Maghimbi and Kazuhiko Sugimura (eds.), *Contemporary Perspectives on African Moral Economy*, pp. 35–54. Dar es Salaam: Dar es Salaam University Press.

Tsuruta, Tadasu 2011. "Africa and Southeast Asia: From the Viewpoint of Comparative Study of Civilization", in Samuel Maghimbi, Isaria N. Kimambo and Kazuhiko Sugimura (eds.), *Comparative Perspectives on Moral Economy: Africa and Southeast Asia*, pp. 91–105. Dar es Salaam: Dar es Salaam University Press.

Turnbull, Colin. M. 1961. *The Forest People*. London: Chatto & Windus.

Ueyama, Shumpei 1962. *Methodologies of Comparative Studies of History*. Tokyo: San'ichi-shobo (in Japanese).

Ueyama, Shumpei 1966. "On Social Formation", in Jiro Kawakita et al. (eds.) *Man: Anthropological Studies*, pp. 73–99. Tokyo: Chuokoron-sha (in Japanese).

Ueyama, Shumpei 1990. *A History of Japanese Civilization I: Tracing the History of Acceptance and Creation*. Tokyo: Kadokawa-shoten (in Japanese).

Umehara, Hiromitsu 1992. *Structure and Change of Rural Philippines*. Tokyo: Kokon-syoin (in Japanese).

Umesao, Tadao 1984. "Keynote Address: Japanese Civilization in the Modern World", *Senri Ethnological Studies*, no. 16, pp. 1–15.

Umesao, Tadao 1989. *Collected Works of Tadao Umesao V*. Tokyo: Chuokoron-sha (in Japanese).

Umesao, Tadao 2003. *An Ecological View of History: Japanese Civilization in the World Context*. Melbourne, VIC: Transpacific Press.

van der Ploeg, Jan Douwe 2008. *The New Peasantries: Struggles for Autonomy and Sustainability in an Era of Empire and Globalization*. London: Earthscan.

van der Ploeg, Jan Douwe 2018. *The New Peasantries: Rural Development in Times of Globalization*. London: Routledge.

von Freyhold, Michaela 1979. *Ujamaa Villages in Tanzania: Analysis of a Socialist Experiment*. London: Heinemann.

Wai, Dunstan 1978. "Sources of Communal Conflicts and Secessionist Politics in Africa", *Ethnic and Racial Studies*, vol. 1, no. 3, pp. 286–305.

Watts, Michael J. 1983. "The Political Economy of Climatic Hazards: A Village Perspective on Drought and Peasant Economy in a Semi-Arid Regaon of West Africa", *Cahiers d'Études Africaines*, vol. 23, nos 89–90, pp. 37–72.

Weber, Max 1968. *Economy and Society: An Outline of Interpretative Sociology* Vol. 1. New York: Bedminster Press.

Whyte, William H. 1958. *The Organization Man*. New York: Simon & Schuster.

Wilkie, David S. and Bryan Curran 1993. "Historical Trends in Forager and Farmer Exchange in the Ituri Rain Forest of Northeastern Zaire", *Human Ecology*, vol. 21, no. 4, pp. 389–417.

Williamson, Oliver E. 1985. *The Economic Institutions of Capitalism*. New York: The Free Press.

Wilmsen, Edwin N. 1983. "The Ecology of Illusion: Anthropological Foraging in the Kalahari", *Reviews in Anthropology*, vol. 10, no. 1, pp. 9–20.

Wilmsen, Edwin N. 1989. *Land Filled with Flies: A Political Economy of the Kalahari*. Chicago, IL: University of Chicago Press.

Wittfogel, Karl A. 1957. *Oriental Despotism*. New Haven, CT: Yale University Press.

Wolf, Eric 1966. *Peasants*. Englewood Cliffs, NJ: Prentice Hall.

Woodburn, James. 1982. "Egalitarian Societies", *Man (New Series)*, vol. 17, no. 3, pp. 431–51.

World Bank 1981. *Accelerated Development in Sub-Saharan Africa: An Agenda for Action*. Washington, DC: The World Bank.

World Bank 1989. *From Crisis to Sustainable Growth – Sub-Saharan Africa: A LongTerm Perspective Study*. Washington, DC: The World Bank.

Yasuoka, Hirokazu. 2011. *Ecological Anthropology of the Baka Hunter-Gatherers: Rethinking Hunter-Gatherer Subsistence in the African Rainforests*. Kyoto: The Center for African Area Studies, Kyoto University (in Japanese).

Yasuoka, Hirokazu. 2012. "Fledging Agriculturalists? Rethinking the Adoption of Cultivation by the Baka Hunter-Gatherers", *African Study Monographs*, Supp. Issue 43, pp. 85–114.

Yatsuka, Haruna. 2012. *Changes in Natural Resource Use and Social Relationships among the Sandawe in Tanzania: Implications with Regard to the Modernization of Hunter-Gatherer Societies*. Kyoto: The Center for African Area Studies, Kyoto University (in Japanese).

Yatsuka, Haruna. 2016. "Historical Interaction with Neighbors from the View of Livelihood Change: A Study of the Sandawe of Tanzania", *Senri Ethnological Studies*, vol. 94, pp. 81–105.

Yatsuka, Haruna 2017. "Attitude of the Hadza Hunter-Gatherers toward Tourism in Tanzania: Focus on Relationship with Neighbours, Individual Mobility and Cash Incomes", *Journal of African Studies*, vol. 92, pp. 27–41 (in Japanese).

Young, Crawford T. 1994. *The African Colonial State in Comparative Perspective*. New Haven, CT: Yale University Press.

Zolberg, Aristide R. 1967. *Creating Political Order: The Party-States of West Africa*. New York: Rand McNally & Company.

Index

Page numbers in **bold** denote tables, those in *italics* denote figures.